# EASTERN LONG ISLAND GEOLOGY

HISTORY, PROCESSES AND FIELD TRIPS

BY

LES SIRKIN



Published by: BOOK AND TACKLE SHOP
7 Bay Street
Watch Hill, RI 02891
401 596 0700

Sole Distributor:

THE BOOK & TACKLE SHOP
Established 1953

7 Bay Street
Watch Hill, R.I. 02891
401 596 0700

ISBN Number 0-910258-18-X

Series Editor: William K. Healy
Cover Design and Field Trip Review: Judy FitzGibbon

Coastal Geology Series by Les Sirkin

Block Island Geology
Eastern Long Island Geology
Western Long Island Geology (in preparation)

EASTERN LONG ISLAND GEOLOGY: HISTORY, PROCESSES AND FIELD TRIPS

Table of Contents, Page

List of Figures, Page

# GEOLOGY OF EASTERN LONG ISLAND: HISTORY AND PROCESSES

## INTRODUCTION

The dynamics of the atmosphere and the ocean have contributed greatly to shaping the Long Island coast; their combined energies have been most effective during great storms when the wind and the sea rise, and waves surge across the beaches. In recent years, several major storms have increased the rate of coastal erosion in southeastern Long Island. Headlands, like Montauk Point, and barrier beaches in West Hampton Dunes and Fire Island have been battered and thinned. Familiar landmarks have been left in ruins, scores of beach homes washed away, and valuable forests destroyed.

In contrast, most geological processes are set in a different time frame. They appear to be much less active in the short-term. The erosive forces of wind and waves seem to be of little consequence only a few hours after a storm. Bluffs retain their steep profiles, even though their slopes are deeply gullied. Beaches also appear to be intact, but with step-like terraces tracing their seaward margins. If the storm is a hurricane and the beach in question suffers a direct hit, then the scale of damage is multiplied by the whole catalogue of destructive phenomena.

In the perspective of geologic processes and time, eastern Long Island has changed dramatically since twenty-two thousand years ago when the Laurentide ice sheet, named for the source region of the glacier in eastern Canada, covered eastern North America between Canada and southern New York. Because of this glaciation, eastern Long Island formed like a slab of icing on top of a geologic layer cake. Long Island is an elongate mass of glacial sediment sitting on layers of clay and sand, all of which cover a basement complex of crystalline bedrock.

First, the geology of eastern Long Island may be considered in terms of the history of the earth's crust and the formation of the continents, a history beginning with the bedrock foundation of northeastern North America. This part of the continent began to evolve in the Proterozoic Era over one billion years ago (Figure 1). At that time, the precambrian shield, the most ancient core of the emerging North American continent, which in places dates to nearly four billion years, collided with another, equally ancient, crustal fragment- perhaps the protocontinent Baltica, another precambrian shield and the ancestor of the European continent.

The rocks that formed the margins of these continents were crumpled and metamorphosed, and a massive mountain range was created between what is now eastern Canada and northwestern Mexico. With an age of deformation of about one billion years, these metamorphic rocks are the oldest in the southern New York and New England upland. This collision was followed in the late Proterozoic by an interval of rifting apart of the new continent- somehow the earth forces that caused the collision reversed gears and the land mass separated. In the process, the sea filled the new ocean basin, and a thick layer of sediment was deposited on the margin of North America.

The next part of this history traces the growth of the Appalachian Mountains during the Paleozoic Era from nearly six hundred million years ago to about two hundred fifty million years ago. These mountains were also the result of continental collisions, a number of which occurred over this interval. This time the crustal plates of ancestral North America, Europe and Africa ground against each other and crumpled the sedimentary continental margins into

| ERAS | PERIODS | AGE (my*) | EVENTS |
|---|---|---|---|
| Cenozoic | Quaternary | 0 to 2 | Cyclic glaciation (see Figure 10) |
| | Tertiary | 2 to 65 | Northern Hemisphere glaciation; Erosion of S.New England upland. |
| Mesozoic | Cretaceous | 65-144 | Continents move toward modern positions;Buildup of continental shelf. |
| | Jurassic | 144-213 | Pangaea begins to split apart 200my;Basins form. |
| | Triassic | 213-248 | |
| Paleozoic | Permian | 248-286 | Pangaea Supercontinent completed;Appalachian Mts completed. |
| | Pennsylvanian | 286-320 | Coal swamps form |
| | Mississippian | 320-360 | |
| | Devonian | 360-408 | Laurasia and Gondwana begin to merge. |
| | Silurian | 408-438 | |
| | Ordovician | 438-505 | Laurasian continents begin to merge. |
| | Cambrian | 505-590 | Avalonia joins N.A. |
| Proterozoic | | 590-2500 | Protocontinents develop |
| Archean | | 2500-4600 | Beginning of Planet Earth,4600my |

Figure 1. The Geologic Time Scale (0 to 4.6 billion years ago; my* = ages in million years). For Quaternary Time Scale see Figure 10).

mountain ranges (Figure 2).

The collision process reversed again during the Mesozoic Era beginning around two hundred million years ago when the supercontinent, Pangaea, made up of all the existing continents, all having merged during Proterozoic and Paleozoic time, began to break apart (Figure 3). Now the Atlantic Ocean basin formed in the widening rift, and sediments from the erosion of the Paleozoic mountain ranges were accreted to the margin of the North American continent. By the end of the Cretaceous Period around sixty-five million years ago, a wedge of sediment had been deposited seaward from the foothills of the Appalachians to the edge of the continental shelf. This mass of stratified sediments rests unconformably on the crystalline rock of the Paleozoic basement (Figure 4).

How Long Island got on top of the pile makes up the third component of this story. The glaciers of the Pleistocene Epoch were responsible for transporting and depositing a vast quantity of sediment derived from the New York and New England upland to form a narrow zone of hills, known as moraines, at the outer margin of the ice. The terminal moraine, the outermost deposits, and several recessional moraines- those formed as the ice melted- of two glaciations now traverse Long Island, Block Island, Martha's Vineyard and Nantucket (Figure 5). In eastern Long Island, moraines form a complex of linear hills and intervening zones of sorted and stratified glacial sediment called outwash. Twice during the last two hundred thousand years, continental glaciers expanded southward from eastern Canada and deposited moraines along virtually the same terrain. The last ice, which arrived some twenty-two thousand years ago, created the rough stock from which the offshore islands would be carved by subsequent geologic forces. More specifically, we now understand the "moraine" that

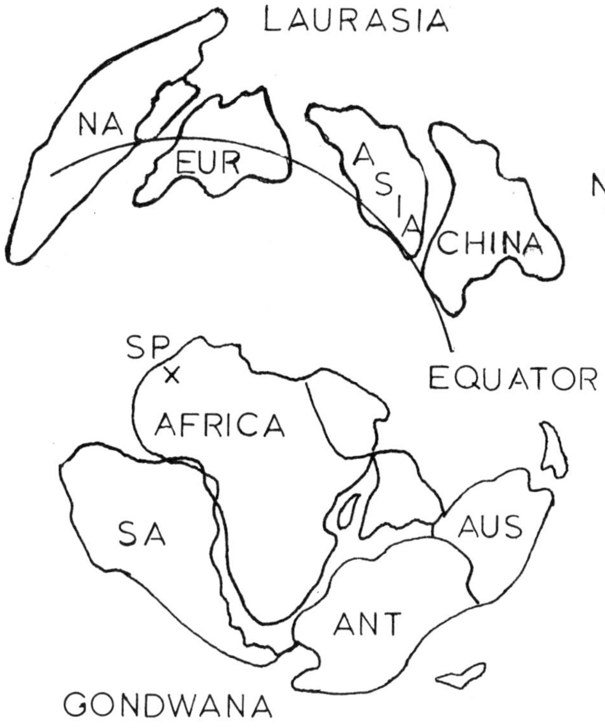


Figure 2. The mid-Paleozoic Continents, Equator and Poles

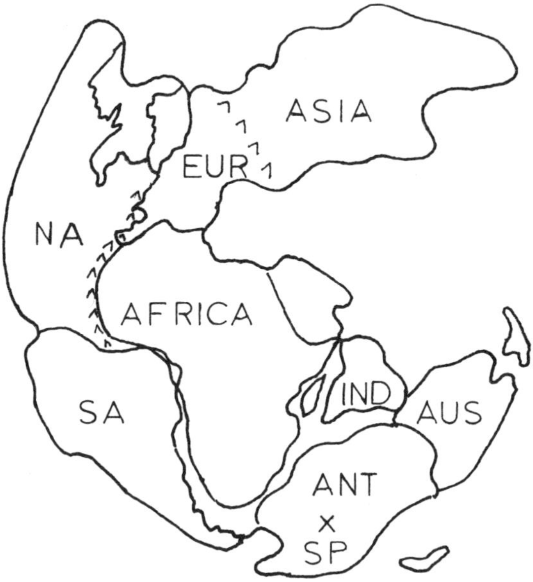


Figure 3. Pangaea

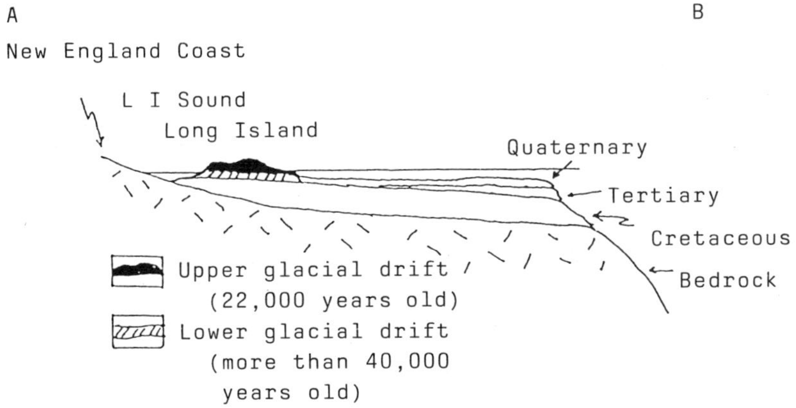


Figure 4. Cross Section of the Continental Shelf off Long Island

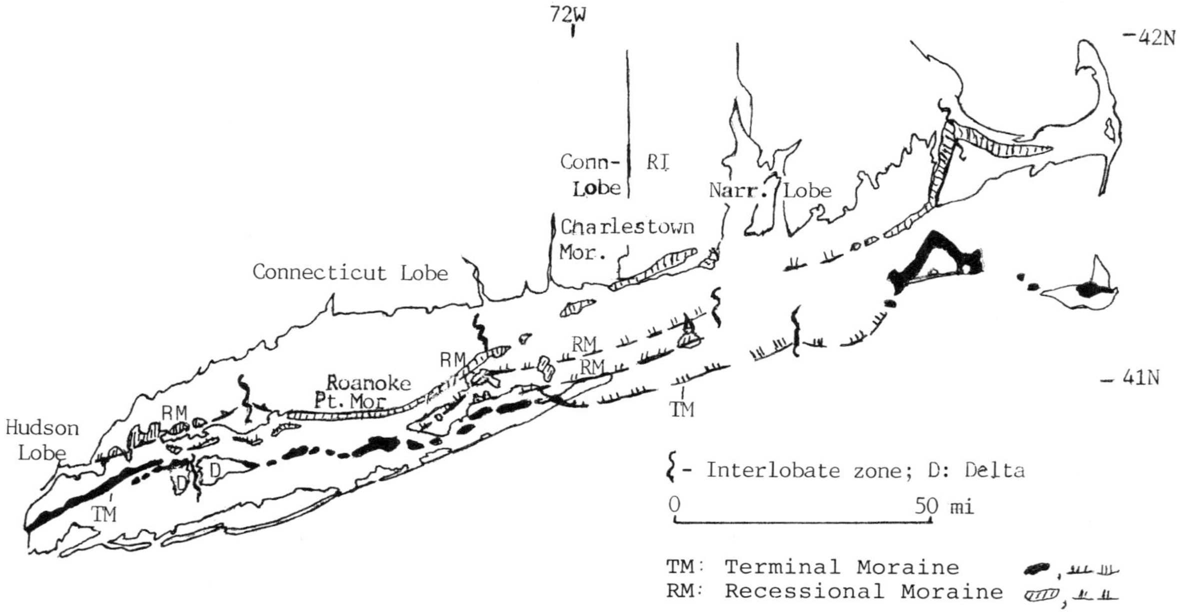


Figure 5. The Moraines of Long Island and Southern New England

dominates the topography of the South Fork is the product of two glaciations and the interaction of at least two glacial lobes that existed during both glaciations.

The last part of this history deals with the transformation of the moraines by on-going geologic processes: the erosive action of ocean currents and rising sea level, as well as by wind, rainfall and running water. All of these erosional forces have contributed to shaping eastern Long Island into its characteristic two forks dominated by glacial topography, straight line barrier beaches and myriad bays, headlands and sand bars. The glaciations and postglacial geological processes make up the most dramatic chapters because they created the land as it is today; the changes imposed by these processes are first-hand evidence of their powers. The Laurentide ice sheet covered southern New England and New York. The ice tapered southward to the outermost range of glacial deposits, the terminal moraine, which today makes up the hilly region of central Long Island and the South Fork. The eroded remains of the moraine are found on the sea bed south of the Montauk Peninsula. The shoreline of the Atlantic Ocean at the height of the glaciation was about seventy miles south of the ice front and near the edge of the continental shelf. The sea, which supplied most of the moisture to the glaciers through evaporation, was over three hundred fifty feet below its present level. Rivers cut canyons into the shelf where ice-age mammals grazed on land covered with tundra and spruce forests (Figure 6). The terminal moraine stood as a massive deposit of sand and gravel marking the outer limits of the ice (Figure 5). An outwash plain composed of sand and gravel was deposited south of the moraine. Smaller tracts of outwash formed between the terminal moraine and the recessional moraines. In this zone of deposition, meltwater carried sand and clay into proglacial lakes to build deltas. The lakes formed when water ponded

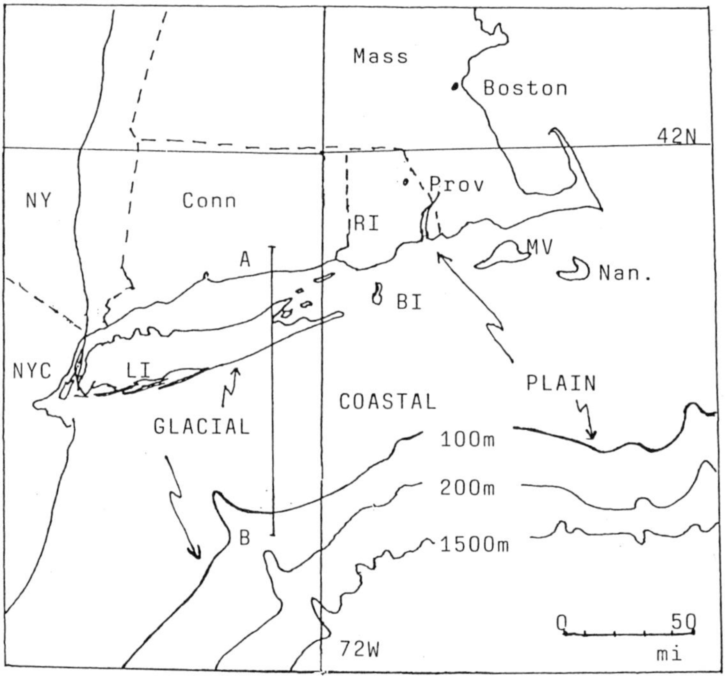


The line A-B locates the cross section in Figure 4.

Figure 6. Southern New England and New York Continental Shelf between the ice and the moraines.

The changes imposed on the surface of the land as the glacier melted and began to replenish the ocean are impressive. If we had been able to see eastern Long Island twelve thousand years ago when the sea was within one hundred feet of its present level, it would have been about three times wider and somewhat longer than it is today, with the recessional moraines above sea level (cover). We would also have seen meltwater accumulating in proglacial lakes between the moraines west of Riverhead, and in the Peconic bays and Long Island Sound. As lake levels rose, water spilled over the lower elevations of the moraine. About this same time ocean currents finally undermined the moraine, and the impounded water sluiced through the gaps and into the sea. Streams rushing from the lakes carved broad and deep channels that exist today as gaps in the moraine and meandering channels in the outwash plain and the sea bed (Figures 5 and 6). Ultimately, the rising sea flooded the lake basins and rose up around the moraines.

Six thousand years ago when the sea approached its present level, eastern Long Island consisted of several east-west ridges separated by fingers of the ocean and outwash plains. From that time onward, the Island has gradually been downsized and streamlined by geologic processes. Coastal currents have connected barrier beaches along the south coast, rejoining some segments of the island previously breached by meltwater streams, and submerging or washing away other tracts of moraine to create the modern geography. Rain, wind and waves have torn away the bluffs, and rainfall has saturated the glacial deposits to create a water table.

More recently, people have imposed their own style of modifying nature, filling in coastal wetlands and eroded barrier beaches. What we see of eastern Long Island today is the product of many processes that, with the rising sea, continue to reshape the land, steadily releasing its treasure of sediment to the encroaching sea, in the face of an embattled population trying to hold on against the natural order. While the geologic story of eastern Long Island is the main subject of this book, some of the contrasts between natural and man-made features are noted. In the final chapter, field excursions are provided to allow direct examination of the geologic evidence and an opportunity to evaluate the changes that have taken place.

# GEOLOGIC HISTORY

## BUILDING THE CONTINENTS

Beginning early in the earth's history, four billion years ago, fragments of the earliest crust were merging to form larger land masses. These fragments included volcanic islands and deposits of rock debris that accumulated in basins on the hardened surface of the planet. Through the Archean and Proterozoic eras, these protocontinents acquired a history of growth in the collisions and separations of crustal blocks of increasing size and complexity (Figures 1 and 2).

Early in the Paleozoic Era, over five hundred million years ago, two major continents, Laurasia and Gondwana, were taking shape and incorporating most of the available continental crust. Laurasia grew from the merging of continental blocks that would eventually become North America, Europe and Asia, along with some large volcanic islands and adjoining basin complexes, features that probably resembled the present islands of Japan and the Sea of Japan. The rocks of one of these island-basin features, known as Avalonia, make up part of eastern New England.

The components of Laurasia collided, beginning in the early Paleozoic and formed mountain ranges along their boundaries. These ranges include the Ural Mountains between Europe and Asia, and the Appalachians between Europe and North America. The suture between ancestral North America and Europe can be found near the New York-Connecticut border. The continental margin where eastern Long Island would eventually form would not begin to emerge for another three hundred million years (Figures 4 and 6). Unlike North America in our modern geography, Laurasia was aligned with the equator running through eastern North America where the Appalachians are today. This region supported tropical climate from what is now Georgia to Northern Canada.

The other early Paleozoic continent, Gondwana, was mainly in the southern Hemisphere and combined crustal elements ancestral to Africa, South America, India, Australia, and Antarctica (Figure 2). Laurasia and Gondwana gradually collided beginning in the Devonian Period around three hundred eighty million years ago and ending in the Permian Period over two hundred fifty million years ago. This amalgamation created the ultimate supercontinent, Pangaea, a fusing of virtually all of the continental crust on earth at the end of the Paleozoic Era (Figure 3). Pangaea remained intact until the end of the Triassic Period of the Mesozoic Era, two hundred million years ago.

Both continental crust and oceanic crust were involved in these collisions, and pieces of oceanic crust were trapped between the continents. Usually lighter continental crust overrode heavier oceanic rock which plunged deep into the earth forming a subduction zone (Figure 7). Occasionally slices of oceanic crust were shoved onto the continent in a process known as obduction. The heat generated in these movements melted continental rock above the subduction zone creating masses of molten rock called magma. Under pressure, the magma was forced upward in the continental crust and slowly crystallized into mountain-sized bodies of granitic rock known as batholiths. Later, magma from oceanic crust was injected through the crust to form veins of basalt in the existing bedrock and volcanoes and lava flows at the surface. The older continental crust was altered by the heat and pressure of the molten

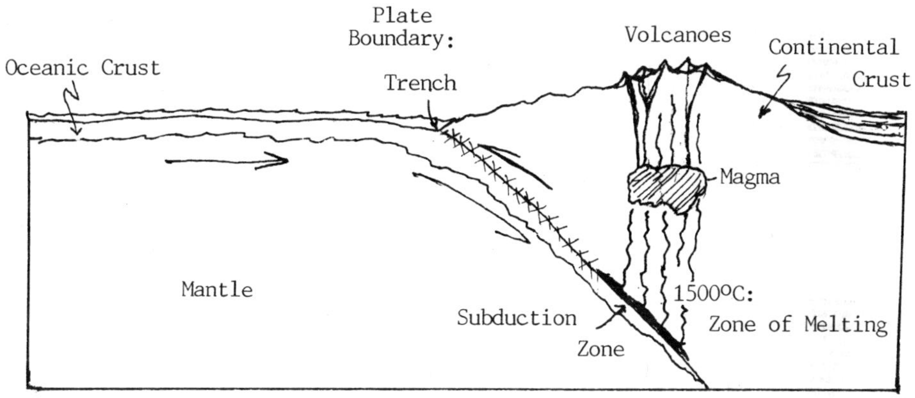


Figure 7. Cross Section of a Subduction Zone

intrusions into crystalline metamorphic rock, known as gneiss and schist. These rocks contain a mix of minerals arranged in light and dark bands that reveal their continental and oceanic affinities.

The modern interactions between segments of the earth's crust that include Australia, India, Asia and the volcanic arcs and basins of the Pacific provide models for the Paleozoic crustal collisions. The collision-in-progress between Australia and Asia is an approximation of the collision that involved Gondwana and Laurasia. Similarly, the merging of India and Asia during the Cenozoic Era has resulted in building the Himalayan Mountains from the marine sedimentary rocks compressed between these continents. This is a modern analog of the late stage in the Paleozoic collisions that formed the Appalachians. From these models we can understand the development of the continents that now border the Atlantic Ocean, and we can reconstruct the collisions and compression of continental and oceanic crust and the thick sedimentary layers that created the mountain ranges of the Paleozoic Era.

Today, southern New York and New England contain the roots of these mountains and include pieces of Eurasia and Gondwana left over after Pangaea broke up. Granitic intrusions fringe southern Connecticut, and gneisses and schists trend northeasterly into New England. In addition, intercalated bands of darker basaltic rocks may represent oceanic crust caught between the continents.

As Pangaea neared completion late in the Paleozoic Era, an elongated embayment formed where the Narragansett Bay lowland now lies. An arm of the late Paleozoic ocean occupied this trough, and coarse gravels were washed into the bay by rivers from nearby mountains. These gravels were soon converted to conglomerate, a type of sedimentary rock made of cemented rock debris. Crustal forces later changed these rocks into metaconglomerates in which the pebbles were stretched and the sandy matrix crystallized. In swamps around the quiet waters of the bay, the remains of forests of primitive trees were ultimately transformed into beds of coal. The southern New York region, including the present position of Long Island, was buried under thousands of feet of sedimentary rock.

At the end of the Paleozoic, Pangaea stood firm with the Appalachians as its backbone. This landmass remained intact for tens of millions of years, well into the early Mesozoic. As the Triassic Period ended, two hundred million years ago, Pangaea began to split apart, literally at its Paleozoic seams. Sheets of lava were extruded from rifts that opened like torn cloth at the edges of the new continents. These events appear to be similar to the rifting that is underway in the east African rift region today. Deep valleys, known as Mesozoic basins, formed in the rifted depressions of eastern North America. Streams eroded the nearby uplands and carried oxidized sediments into the basins. These sediments would become the red sandstone layers over which black basaltic lavas would flow. This continent-splitting cataclysm created several basins, including the Hartford Basin which is now occupied by the Connecticut River Valley (Figure 8). Here, red and brown sandstones are capped by basaltic lava flows that form cliffs along the valley. Other rifts are buried in the continental shelf, and one of these lies south of Long Island. Another, the Newark Basin of eastern New Jersey, is wedged between the Appalachians and the coastal plain. The breaking up of the earth's crust into large blocks, or plates, and movement of these plates, made possible by the softening of the rock beneath the continents, coupled with the circular convective flow of heat within the earth, are now believed to be responsible for both the collision of continents and their eventual

a)

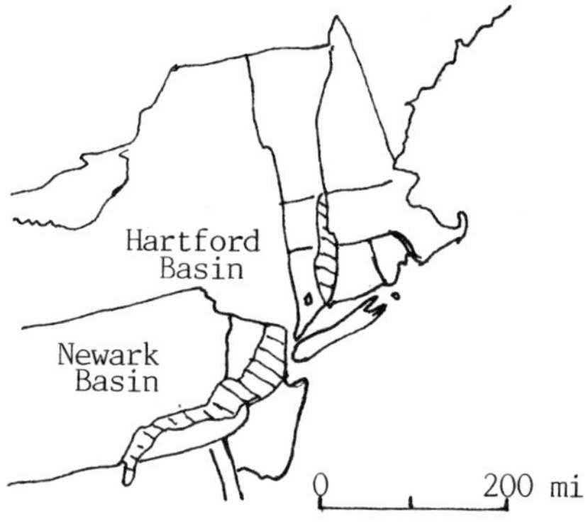


b)

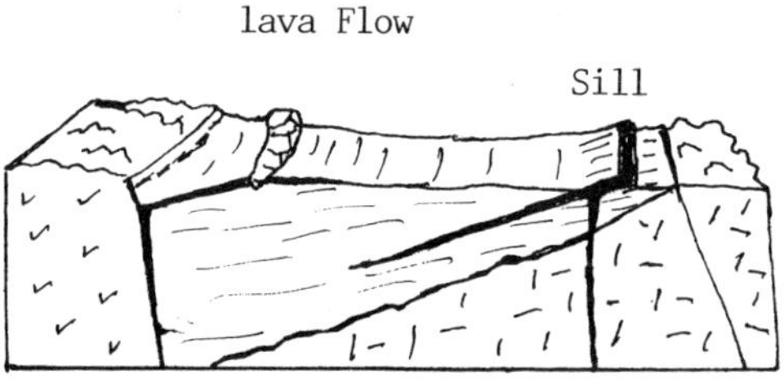


Figure 8. Northeastern United States Generalized Mesozoic Basin and Cross Section

separation. Geologists refer to these processes collectively as Plate Tectonics.

Gradually, the Atlantic Ocean filled the widening abyss between the separating continents. As they moved slowly apart, the plates with their continents realigned themselves geographically (Figure 9). North America rotated counterclockwise from an equatorial position to a north-south orientation, and South America, Antarctica and Australia all broke away from Africa. The rocks that were deposited on the margin of the North American plate in equatorial seas over two hundred fifty million years ago during the Paleozoic were carried toward the north polar region. This juxtaposition of tropical coal beds and coral reefs within a modern polar desert seemed incongruous to geologists of the early twentieth century. It was not until the 1960's that plate tectonics as a working hypothesis unlocked these mysteries.

As the sea floor widened and the continents moved apart, the sea flooded the continental lowlands, and sediments eroded from the continents were piled layer upon layer on the new shallow sea bed. Between the Jurassic Period and the present, sediment layers accumulated to form a wedge of coastal plain and continental shelf sediments thousands of feet thick. The shelf south of the present southern New England coast includes nearly one thousand feet of Upper Cretaceous deposits beneath southeastern Long Island (Figure 4).

During the Tertiary Period between sixty five million and two million years ago, layers of marine sediments were deposited well to the east and south of eastern Long Island. Glacially thrusted beds of Tertiary marine strata that contain fossils of fish and marine mammals wound up in the terminal moraine in Martha's Vineyard. But Tertiary beds were not deposited in the Long Island shelf sequences which were probably above sea level at that time. Eastern Long Island does have late Cretaceous sands and clays, some of which contain thin seams of lignite,

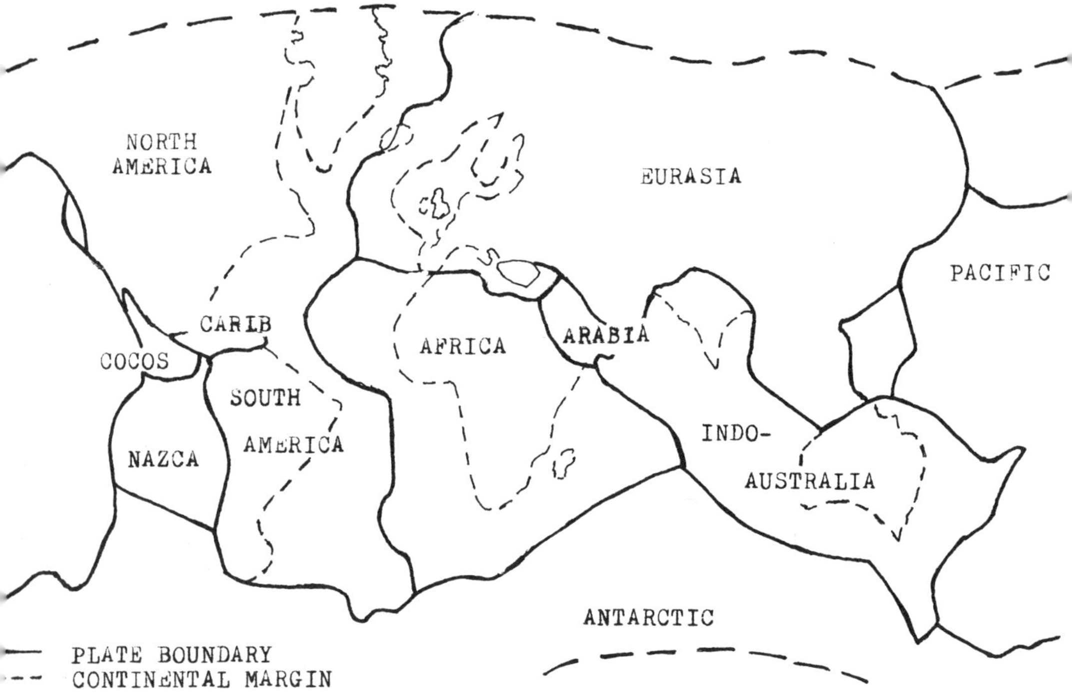


Figure 9. A Generalized Map of Crustal Plates

a type of soft coal, and nodules of iron pyrite- called fools' gold- that formed in ancient wetlands on the margin of the Cretaceous sea. Like the Tertiary of Martha's Vineyard, some Cretaceous sediments have also been thrusted into the moraines, transported southward by the glacier from their sea-floor outcroppings beneath Long Island Sound. They now reside as disoriented blocks stuck in the moraines in the mix of rock debris derived from the Paleozoic bedrock of the mainland.

The boulders on the beach are examples of glacial erratics, pieces of bedrock that are out of context in their present locations. These rocks are a reminder of Paleozoic plate tectonics and contain minerals that reveal their origins deep in the crust. They call to mind the collisions between oceanic and continental plates, rifting of the crust, and the intrusion of igneous magmas. With these clues in hand, we can envision crustal plates colliding and volcanic islands off the coast of ancestral North America. Granite boulders may suggest ancient batholiths, while green and black cobbles could be derived from oceanic lavas. Weathering and erosion of thousands of feet of bedrock since Paleozoic time has exposed the deeper crust of the mainland to the work of glaciers that, in turn, brought this material to eastern Long Island. Conveniently, these rocks from differing northern terranes help to identify the source, or provenance, of the glacial debris.

## GLACIATIONS

In the sixty-three million year interval of the Tertiary Period, the Cretaceous strata of the coastal plain stood above sea level. The shoreline was tens of miles south and east of Long Island, and rivers from the mountains dissected the plain, excavated valleys, and carved rolling hills into the upland. The coastline as we know it, from eastern Long Island to Cape Cod, had yet to emerge in this ancient geography. We can only infer from the shape of the sea floor what the land might have looked like before the last glacier arrived (Figure 6).

It is probable when glaciations began early in the Pleistocene Epoch, the advancing ice flowed like a river, filling the lowlands, thinning over the hills, and extending out of the major valleys as lobes of debris-ladened ice (Figure 10). Thus the shape of the land controlled the flow of the ice during glaciations. Major landforms like the Narragansett Bay lowland, the Connecticut-Rhode Island upland, the Connecticut River valley, and the Hudson River valley channeled the ice into south-trending lobes that directed ice flow and oriented rock fragments (Figure 5). Early in the 20th century, many geologists believed there had been four Pleistocene glaciations in North America. The sediments of each of the glaciations and the intervening warm intervals, the interglacials, were assumed to have been universally deposited like the layers of a cake. Today we know there were more than sixteen Pleistocene glaciations during a few major glacial stages. The deposits of most glaciations were limited in geographic extent, and very few glaciers reached the maximum glacial limit. In most glaciated regions, thousands of years of erosion separated glaciations, and advancing ice removed much of the older

| PERIOD | EPOCHS | STAGES | AGE yrs | EVENTS |
| --- | --- | --- | --- | --- |
| Quaternary | Holocene | Recent | 0-10,000 | N and S BI joined, sea level rise |
| | | Late | 10,000- | Sea Level Rise; Glacial lakes drain 12,000 |
| | | | 20,000 | BI ice free Glacier retreats |
| | Pleistocene | Wisconsinan | 22,000 | Glaciation; low sea level |
| | | Mid-Wisconsinan | 30,000 | Warm Interval Sea level rises |
| | | Early Wisconsinan | 60,000 | Glaciation; low sea level |
| | | Sangamon | 125,000 | Interglacial; High sea level |
| | | Illinoian | 140,000-200,000 | Early (first) LI glaciation (?) |
| | | Early Stages | 200,000-2,000,000 | Interglacials & Glacials |

Figure 10. Quaternary Period Time Scale (0 to 2 million years ago)

glacial sediment.

For eastern Long Island, there is evidence of only two glaciations. The most recent glacier, the Laurentide Ice Sheet, reached its greatest extent twenty-two thousand years ago, late in the Pleistocene Epoch, in the latter part of the Wisconsinan Stage. The earlier glaciation took place before forty thousand years ago either in the early part of the Wisconsinan Stage, or more than one hundred forty thousand years ago in the next older glacial stage, the Illinoian.

Field studies of glacial deposits in southern New York and New England, and particularly in Long Island and Block Island, show the existence of two distinct layers of glacial deposits, known as drift sheets. These two drift sheets were the products of the two glaciations. The lower, or older, layer was deposited by the first glacier. In easternmost Long Island, this drift was derived from a glacial lobe that traversed the Narragansett lowland from the northeast and extended southwesterly as far as the South Fork of Long Island. The upper, or younger, drift sheet was deposited from a lobe of the second glacier. This ice crossed from southern New England to the South Fork from northwest to southeast. In east central Long Island the flow spread southward forming a lobe directed by the Connecticut valley and flowing southeastward on its eastern margin and southwestward on its western margin.

A key element in the understanding of this two-part glacial history has been the recognition of the extensive rearrangement of pre-existing sediments by the last glacier. Also of considerable importance is the attempt to sort out the ages of the early drift sheet and the sedimentary deposits that were formed during the intervening warm climatic intervals: the Sangamon Interglacial, the interglacial that separates the Illinoian and Wisconsinan glacials; and the mid-Wisconsinan interstadial, referred to as the Portwashingtonian on Long Island, a warm interval within the Wisconsinan glaciation.

## WARM INTERVALS AND THE AGE OF THE LOWER DRIFT

The older of the two drifts predates the limit of radiocarbon dating, which is about forty thousand years. We can propose a relative age for the glaciation based on the stratigraphic position of its drift, that is, it is found below the upper drift. Also, it is somewhat weathered, and masses of lower drift are incorporated in the upper drift. But in order to establish a finite age for this glacial event, another radioactive element and a host mineral must be found. The most likely element is uranium, which is the basis for uranium-series dating. Uranium is found in crystals in igneous rocks and the shells of marine animals that extract uranium from sea water. Because no igneous rocks of late Pleistocene age were formed and interbedded in the drift in this region, the crystals in marine shells could supply the answer.

But how might marine animals have entered the glacial-interglacial record? If the interglacial between the first and second glaciations was significantly warm, and sea level rose high enough, warm-water marine animals could have lived in the sea where Long Island Sound is today. Coral, a mainly tropical, marine animal, has been widely used for dating the last interglacial and providing an estimate of sea level at that time. Fossil coral found in undeformed marine deposits would provide an age for the sediments and the warm episode in question. If the coral was picked up by a glacier and redeposited in the moraine, the age of the coral would be that of the warm interval during which it grew; but it would only be a minimum age for the older glaciation. In Nantucket, a fragment of coral redeposited in a sand layer in the Sankaty Cliffs was dated at about one hundred thirty thousand years. While not a finite age for the sand unit, the coral provides an age for the interglacial event. Thus far, however, coral has not been found in the Long Island drift sheets.

The geological record of this part of the Atlantic coastal plain shows at least two warm intervals in the late Pleistocene. The older, the Sangamon Interglacial, peaked around one hundred twenty five thousand years ago (Figure 10). It separates the Illinoian and Wisconsinan glacial stages, and it was warmer and had a sea level about ten feet higher than today. In the classic study of the geology of Long Island, published in 1914 by United States Geological Survey, the author, Myron Fuller, describes a marine sedimentary unit, the Gardiners Clay, named for the site of its discovery on Gardiners Island, as a marine clay deposited during an early Pleistocene interglacial. A number of other clay beds, described by Fuller and later authors, were also relegated to this catch-all unit, the "Gardiners Clay." More recently, the Gardiners Clay of Gardiners Island has been shown to be an isolated, glacially-thrusted mass of marine sediment whose age and provenance are uncertain. Its fossils indicate a colder water origin than that expected in an interglacial deposit at this latitude. Many of the other clay beds that crop out in the morainal zone and were previously correlated with the "Gardiners" have turned out to be freshwater lake clays.

There are also marine clay beds in place in the subsurface of southcentral Long Island, south of the moraines. These are also called the Gardiners Clay, and they are also of uncertain age. Marine clays have been encountered from as little as two feet to as much as two hundred feet below the surface. At Brookhaven a clay identified as the Gardiners Clay is around ninety feet down and seems to represent a time, based on its pollen content, when the climate was as warm as the present. Unfortunately, none of the Gardiners Clay samples have provided corals for uranium-series dating, so that the age of the subsurface unit is still unknown. Subsurface clays, like the Gardiners, may be important hydrological units that form confining layers for the region's aquifers.

The younger, warm episode represents a mid-Wisconsinan interstadial centered around thirty thousand years ago. This event was not as warm as the Sangamon, and the sea was about sixty feet shy of its modern level. In Long Island, the warmest interval of the mid-Wisconsinan interstadial is dated between thirty thousand and twenty-eight thousand years in both undisturbed marine strata off the south shore and glacially-transported marine sediments. During the last glacial advance, reefs of oyster shells, salt marsh peat and clay were excavated by the ice from the floor of a marine embayment north of Long Island and then redeposited in the terminal moraine. This sedimentary sequence, the Portwashingtonian beds, named for their location in western Long Island, includes a variety of marine and glacial sediments, all radiocarbon dated between more than forty thousand years and twenty-one thousand seven hundred fifty years. Wood found in glacial till interbedded with the organic-rich sediments provides an age for the arrival of the Laurentide glacier on Long Island twenty-two thousand years ago.

The interstadial Portwashingtonian beds, located both physically and stratigraphically between the tills of the older and younger glaciations, establish a minimum age of forty thousand years for the older deposits. Through their fossil content, mainly the oysters and pollen of evergreen and hardwood trees, these beds also provide us with a glimpse of vegetational response to changing climate, as well as coastal ecology, during that interval.

By extrapolating the Portwashingtonian and Sankaty ages to eastern Long island where theoretically there are matching sediments, we can infer that the older drift sheet was deposited before forty thousand years, either during a glacial advance about sixty thousand years ago in early Wisconsinan time, or during the Illinoian glaciation more than one hundred forty thousand years ago. The younger age would be in keeping with the sequence of ages and climatic fluctuations in the late Pleistocene of Long Island. For example, fossil spruce pollen in the Portwashingtonian beds indicate cold conditions around forty thousand years ago at the onset of a warming trend following an earlier glaciation. And at Montauk, a thirty eight thousand year age has been determined for sediment containing organic compounds found between the older and younger drifts.

Support for an Illinoian age for the earlier glaciation is based on the distribution of glacial deposits and the inferred geographic extent of the two earlier ice sheets. Some geologists limit the early Wisconsinan glaciation to the Lake Ontario and St. Lawrence region. Furthermore, morainal deposits found south of the Wisconsinan terminal moraine in eastern Pennsylvania are believed to be of Illinoian age. These determinations make an Illinoian age for the early glaciation on Long Island worth considering.

## THE LOWER, OLDER DRIFT

Both the Sangamon Interglacial and the mid-Wisconsinan interstadial were episodes of prolonged weathering and erosion. Each could account for the oxidation of the existing drift and erosion of the coastal headlands. The till of the lower drift often has red, iron-oxide weathering on its joint planes. During both warm intervals, sea level was near its present level. The Sangamon sea may have been higher than modern seas, and its mid-Wisconsinan counterpart was as much as sixty feet lower. The much broader-than present Montauk headland would have been battered by the sea much as it is today.

The lower drift has also been deeply channeled by stream erosion that must have happened during a significantly lower sea level like that of a glacial interval. These channels have been filled variously with marine clay and sand where they were open to postglacial marine transgression, and glacial meltwater gravel and outwash in their upland reaches. In places this drift appears to have been planed off by the late Wisconsinan glacier, while at Montauk Point it rises above the present beach like a moraine. As such, the older drift forms the bulwark of the bluffs where it is dramatically exposed. It often appears to be dark gray in color, crudely stratified, and with lenses of sand and beds of lake clay and silt. Where exposed to the spray of breaking waves it becomes case hardened with salt rinds and takes on a steely gray appearance. This drift is also deformed, contorted into tight folds, ripped into large blocks and thrusted southward. Many of the ripped-up blocks are surrounded by the light-colored outwash and deltaic sands of the upper drift. These intense alterations seem to have been imposed during the southward advance of the younger glacier which filled in behind the folds and thrusts with sand and gravel washed out of the melting ice.

The older drift also contains compact basal till, an unsorted and unstratified glacial sediment that was deposited beneath the ice sheet. In an undisturbed condition, this till is relatively impermeable and resists infiltration by soil water. While the till has a low proportion of clay to its sand and gravel content, its compaction accounts for the tightness and the resulting ponding of surface water. As a whole, this drift is continuous beneath the surface layer of younger drift in the Montauk Peninsula and along the northern half of the South Fork. The ripped-up blocks of the drift are seen locally where bluffs are cut into the moraine, but are hard to interpret from subsurface data. The till in the lower drift changes little in character between its eastern and westernmost exposures in the South Fork. At both Montauk Point and Cow Neck, it is a loamy deposit with blocks of bedded clay. These clays, which are rhythmically banded, may have originated as varved clays that were deposited in proglacial lakes.

At the height of the first glaciation, the ice over eastern Long Island may have been a few hundred feet thick, tapering to a terminus several miles to the south. The ice margin formed morainal hills in the South Fork and extended southeastward onto the continental shelf south of the Montauk Peninsula. This stretch of the moraine is now beneath the sea, and it was heavily eroded during the Holocene sea level rise. As the early glacier receded, a lake formed between the ice margin and the terminal moraine, and layers of varved silt and clay accumulated on the lake floor. Later, forward movement of the ice crumpled the lake beds into tight folds. At the end of the early glaciation, the place where eastern Long Island would eventually take shape was buried somewhere in the middle of the terminal moraine of that glacier. An understanding of the deformation of the lower drift, which includes the Montauk Till, has proven to be important to the interpretation of the geology and the hydrology of the South Fork. In this region, the Montauk Till can be the basal till in the superposed sequence of various drift deposits and a lodgement till formed beneath the ice. But it is also a major component of the "ripped up" clasts of till displaced into the younger outwash.

Today we encounter the Montauk Till at or near the surface northward from the southern edge of the moraine in the South Fork to the bays. In the West Neck Road area and Cow Neck, it forms the relatively impermeable platform on which numerous ponds and drainage channels have formed, for example between Cold Spring Pond and Wooley Pond. It is also the surface on which the late Wisconsinan glacial deposits rest (Figure 11). As the cross sections (Figures 11b and 11c) illustrate, north of Bridgehampton the crest of the early Wisconsinan moraine rises to an elevation of one hundred twenty feet above sea level beneath a cover of younger drift,and the Montauk Till thickens to forty feet. At East Hampton the till is over seventy feet thick in the excavation in the local landfill. It has also been exposed in other South Fork excavations, such as the Southampton landfill, and it is encountered in septic system installations and water well borings. It is at or near the surface in North Haven and Sag Harbor, and it underlies late Wisconsinan kame deposits in Bay Point, Redwood, and Barcelona Neck, and along the southeast shore of Northwest Harbor. The till is also revealed in deeper meltwater channels, for example west of the Long Pond-Crooked Pond channel in the Sag Harbor landfill, in the old Easthampton brush dump in the vicinity of Northwest Road, and adjacent to Hands Creek Road, Middle Highway, and Acabonack Road. The lower drift supports bluffs between Ditch Plains and Montauk Point and crops out north of the late

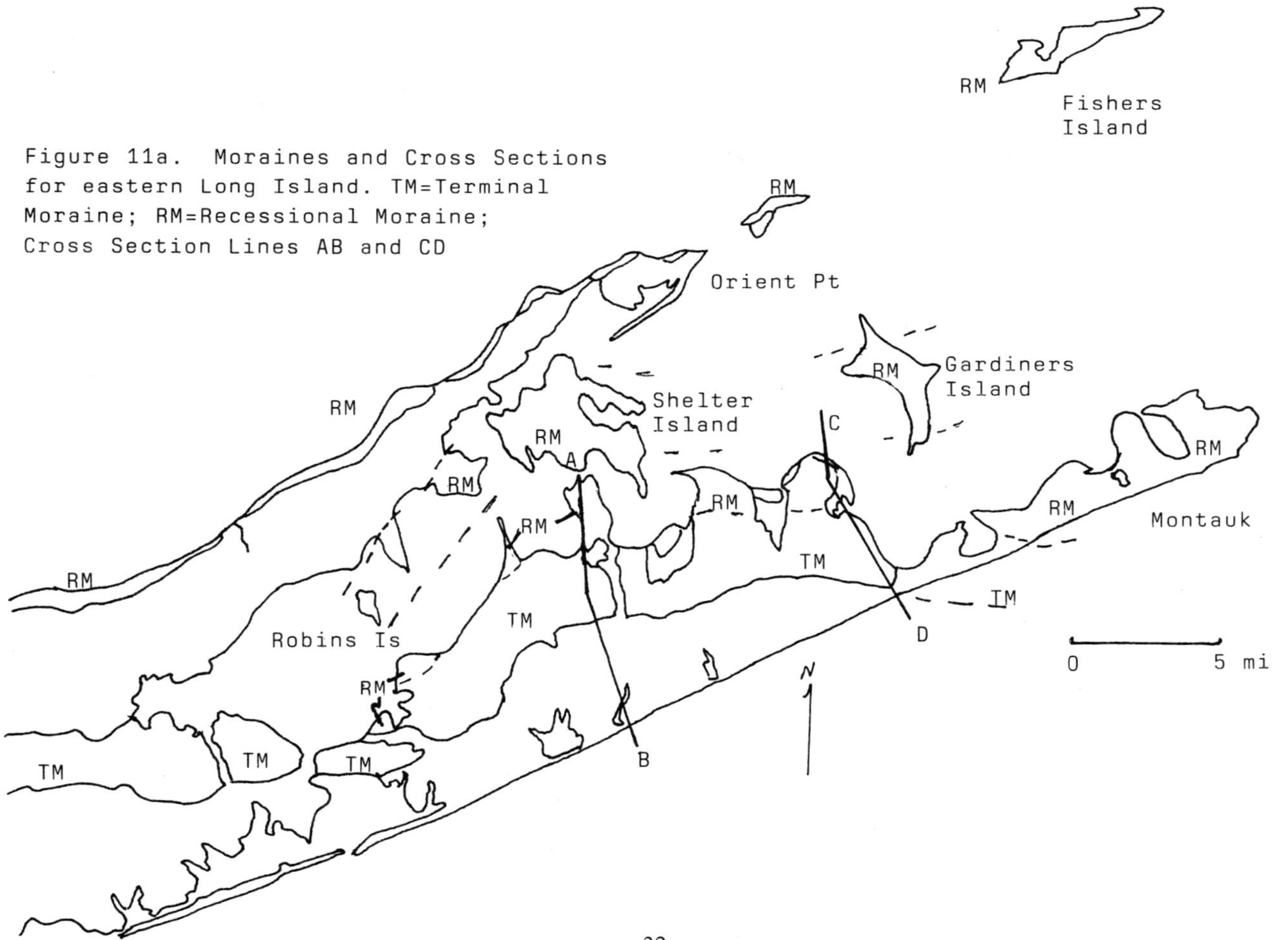


Figure 11a. Moraines and Cross Sections for eastern Long Island. TM=Terminal Moraine; RM=Recessional Moraine; Cross Section Lines AB and CD

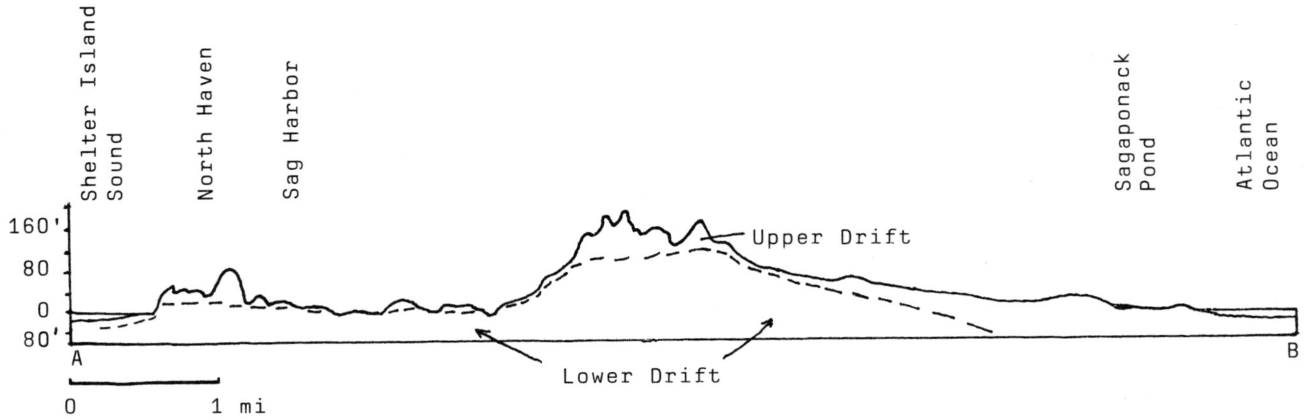


Figure 11b. Cross Section AB: North Haven to the Atlantic Ocean

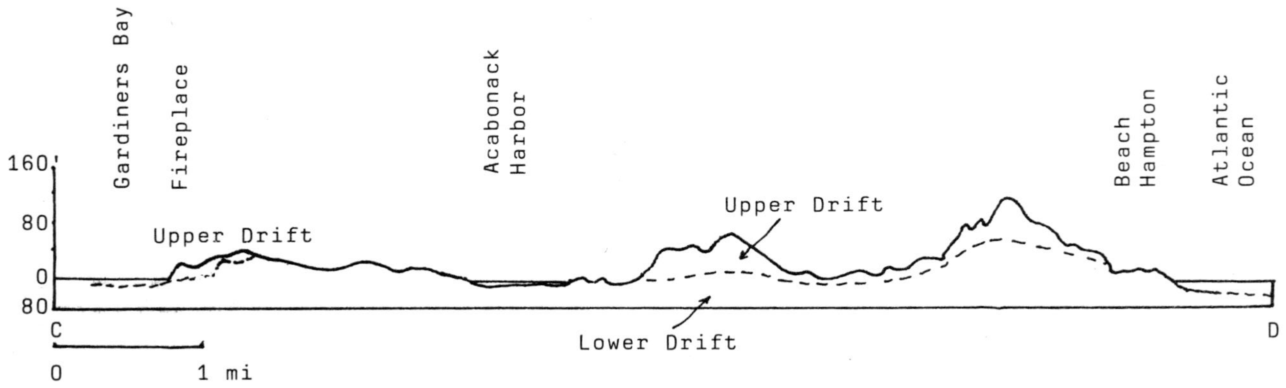


Figure 11c. Cross Section CD: Gardiners Bay to the Atlantic Ocean

Wisconsinan recessional moraine at Rocky Point, Culloden Point, and Shagwong Point.

It has already been noted that the lower drift at Montauk bears some relationship to the geology of the Narragansett embayment of Rhode Island. Till fabrics and till stones at Montauk and Rocky Point show that the early glacier expanded southwesterly across southern Rhode Island toward Block Island and eastern Long Island. While the till fabrics and till stones from the Sag Harbor landfill point to a northeasterly source, till stones from Southampton are aligned in a northerly direction and include Mesozoic basin rock types. Not surprisingly, the red sandstone cobbles at Cow Neck have a Connecticut Valley provenance.

## THE UPPER, YOUNGER DRIFT

The upper drift sheet, the layer of sediments deposited by the late Wisconsinan glacier, overlies the lower, older drift sheet and covers much of eastern Long Island. Much of the younger drift is characterized by glacial deposits and landforms associated with a melting and receding glacier. It contains, in part, a type of till known as meltout till, which as the name implies was melted or washed out of the wasting ice. This till often overlies outwash, which is water laid sediment, also washed from the ice, but sorted in grain size and stratified by the meltwater. The broad expanse of outwash in front of a moraine is an outwash plain. Much of southern Long Island originated as an outwash plain. In eastern Long Island, a segment of the outwash plain is enclosed between the moraines in the Bridgehampton area. This plain is several miles wide and extends southward beneath Shinnecock and Moriches bays and the barrier beaches, where it lies below recent sand deposits on the continental shelf. Where meltwater flows from the ice to the adjacent ground, outwash piles up into conical hills called kames, and when meltwater flows directly from the ice into standing water, the sediment forms a kame delta.

Standing water, as lakes and ponds, is a common feature of a decaying ice front. Lakes often form between the receding glacier and an adjacent moraine. Lake deltas are built by streams flowing into standing water where they form distinctive sets of strata. Sand and gravel collect in the stream channels and at water level as the upper, or topset, beds of the delta. Fine sediment filters down onto the lake floor where layers of silt and clay collect. These are the bottomset beds. More steeply pitched sand layers usually between the topsets and bottomsets become the foreset beds of the delta front. Glacial lakes that undergo freeze and thaw cycles often produce rhythmically bedded bottomsets arranged in silt and clay couplets. These units, termed varves, have been associated with annual cycles of deposition in some lakes. The varves are distinguished both by grain size-mainly clay and silt- and color. In the annual variety, the coarser and lighter-colored, silt layer is deposited when the lake is unfrozen and meltwater carries oxidized sediment into the lake. The darker-colored clays suspended in the still water of the frozen lake gradually settle onto the lake floor to form the other half of the set.

The upper drift sheet contrasts in sedimentary and structural characteristics with the lower drift sheet. The physical deformation of the lower drift by the last ice is impressive, and it includes the younger outwash which was deposited as the ice advanced. Light-colored sands of the upper drift engulf ripped up blocks of lower drift; the outwash overlies deformed lower drift and is often folded. The deformed lower drift and upper outwash are usually covered by apparently undeformed till, often meltout till, which is evidence of the glacier melting and receding. Early on geologists interpreted this structural unconformity: undeformed till over deformed outwash and till as the boundary between the lower drift and the upper drift. They were unaware of the active involvement of the last glacier in deforming both the lower drift and its own outwash, deposited continuously as the ice advanced. Nor did they compare the compositional variations between the lower drift and the upper drift that now help establish the boundary.

When we view the bluffs at Montauk Point, we are looking at a portion of the moraine well north of the glacial terminus. Both the first and the last ice sheets extended south of the

Montauk Peninsula, and considerable glacial sediment has been removed by the rising sea and coastal erosion over the last twelve thousand years. In a fifty foot thick exposure, the Montauk Bluffs reveal remnants of both the older and younger drifts graphically depicting the power of the ice to deform both the younger and older sediments and then, in a change of mood, melt away leaving an upper layer of meltout till. At first it is hard to distinguish the deformed and reorganized lower drift from the upper outwash. However, the light, sand colored outwash seems to surround the fragmented blocks of lower drift. The color difference in the two units is the result of their different provenance, and the till fabric and till stone variety in the upper drift point to a northwesterly source area. Furthermore, it is easy to overlook the similarity in composition of the upper drift outwash and meltout till. This drift is rich in rock debris from the granitic and gneissic bedrock in eastern Connecticut and western Rhode Island over which the last glacier advanced. The ice picked up rock debris ranging from room-sized erratic boulders to cobbles, gravel and sand and covered the lower drift in eastern Long Island with outwash and till.

As the ice sheet melted down, the ice front began to recede, and large quantities of water and sediment were released. Meltwater streaming from the ice cut channels into the drift and laid down a broad outwash plain. Streams coalesced into rivers that flowed across the coastal plain for nearly seventy miles and with a drop in elevation of almost four hundred feet to the shore of the late Wisconsinan sea at the edge of the continental shelf. While meltwater deposited sediment well south of the glacier, moraines of recessional deposits formed when the melting ice margin became stationary for a time. Along the Montauk Peninsula, the crest of the prominent recessional moraine runs east-west through Prospect Hill and Montauk Point, marking a major stillstand of the ice and covering the deformed lower drift. The surface of the so-named Prospect Hill Recessional Moraine segment probably still looks much like it did after the glacier melted.

We can conclude, therefore, that the Upper Wisconsinan drift sheet consists of till and outwash. Locally, the outwash may be as much as two hundred feet thick as it is in the terminal moraine north of Southampton, Bridgehampton and Amagansett (Figure 11). Up to ninety feet of Upper Wisconsinan outwash capped by a thin layer of meltout till that make up the terminal moraine overlie the Lower Wisconsinan moraine nearly doubling the height of this end moraine complex. Undeformed till of varying texture and thickness, usually covers the moraines. The meltout tills are often coarse grained and include rocks, such as granite, granite gneiss, basalt, and red and brown sandstone derived from central and eastern Connecticut. In the recessional moraines, the till may exceed five feet in thickness, as in the recessional moraine at Sebonac Neck, or they may be only a thin veneer of till stones. Meltwater-transported till stones are also scattered over the outwash between the recessional moraines.

Late Wisconsinan deposits also include proglacial lake beds, meltwater channel gravels, and clay and peat in kettle lakes. Masses of silt and clay identified as proglacial lake and delta sediments have also been incorporated in the moraines; one large mass of silt was embedded in the kame at Sebonac Neck. At Montauk, a sandy delta above the glacial strata marks the level of a proglacial lake that formed between the melting ice to the north and the moraine. Channel gravels have been found scattered between morainal segments throughout the South Fork. Subglacial stream channel sediments have been seen in excavations in several of the landfills. Deformation by the last glacier is also apparent in the recessional moraines. At

Hedges Bank, where outwash is capped by till, the upper part of the till has been sheared off and thrust over its till-stone counterpart to the south.

Meltwater was also impounded on the glaciated surface in depressions known as kettles. These lake basins formed in the moraine where large ice blocks were left embedded in the drift as the ice front receded. Where the ice blocks penetrated the impermeable lower drift, water was able to infiltrate the subsurface aquifers, and the kettles remained dry. Elsewhere, the impermeable sediments retained water in the basins and wetlands formed. In one meltwater channel near Scuttlehole, residual ice blocks formed a chain of kettle lakes from north of Short Pond through Long Pond to south of Kellis Pond. The alignment of these kettles mirrors the orientation of the ice mass that formed the channel. Subsequently, water was able to accumulate in the kettles to maintain the lakes. Chains of ponds like the Scuttlehole group have been likened to rosary beads and are known as pater noster lakes. In many of the ponds and lakes of eastern Long Island, fine sediment carried into the basins by surface drainage helped to seal them. In some cases, layers of sediment filled the depressions, which were later transformed into peat bogs and wetlands.

## THE TERMINAL MORAINE

The terminal moraine in the eastern half of Long Island trends in an easterly direction beginning on the east side of a series of north-south ridges that lie south of Huntington and include Half Hollow Hills, Manetto Hills and Dix Hills (Figures 5). These hills comprise a zone of glacial, fluvial and deltaic sediments laid down at the junction of the Hudson and Connecticut lobes of the Laurentide glacier during the late Wisconsinan. In terms of their glacial origin, they are referred to collectively as the Manetto Hills Interlobate Zone, and their geology includes the usual morainal sediments: outwash and meltout till, plus flow till interbedded in the outwash, deltaic beds of ornately cross-bedded sands (topsets), steeply dipping sands (foreset beds), layers of silty clay and clay (the bottomset lake beds), as well as a variety of gravel deposits in meltwater channels that dissect the hills.

The segment of the terminal moraine between the interlobate zone and Riverhead is the Ronkonkoma Moraine. The Ronkonkoma Moraine has been described as a kame moraine in that it consists of a linear series of individual and coalescing kames which are made largely of outwash and are often cut by thrust structures. Much of the moraine is capped by a layer of sandy till. Near Manorville, the moraine is thrusted and a large mass of silt was emplaced in the outwash. A thin seam of shell fragments indicates the silt is of marine origin, but the shells are too weathered for radiometric dating. This exposure occurs in the north-facing re-entrant angle where two sublobes of the ice joined. South of this moraine lies the extensive outwash plain that makes up half of the terrain of Long Island. Lake Ronkonkoma, a large kettle lake, that seems to have been scooped out of the middle of the moraine lies near the path of the Nissequogue-Connetquot drainage. The ice that formed the depression may have been abandoned in a meltwater channel of the last glacier. Adjacent to the lake is its twin kettle, now a wetland obscured by vegetation and filled with sediments that have accumulated since the end of the glaciation. This northerly basin has provided a record of the vegetation since that time as interpreted from the pollen and spores in the sedimentary layers. This history will be described in a later section.

The Ronkonkoma Moraine peaks at over three hundred ten feet at Bald Hill southwest of Riverhead. South of this high point, deposits of upper drift overlie thick lower drift, and eastward an interlobate angle is formed by the Ronkonkoma Moraine to the west and the Shinnecock Hills Moraine to the east. The east-trending Shinnecock Hills segment of the terminal moraine is composed of small kames- generally less than one hundred feet high- between Shinnecock Bay and Great Peconic Bay. From Shinnecock Hills to Tuckahoe the kames lie on a lower drift platform, but many have been extensively mined and have been replaced by buildings and parking lots. As they were gradually stripped away, their contents were shown to include cross-bedded sands, deformed masses of lower till, ice shove or collapse structures and a thin meltout till cover. South of the moraine, the relative position of land and sea obscures the outwash plain which is largely under water; outwash in the wave-cut cliffs along the bay south of the moraine is evidence of its presence.

East of the Shinnecock Hills Moraine, the terminal moraine broadens into a wide band of hummocky, kame, kettle and channeled topography, and increases in height. Along Deerfield Road, the moraine is about three miles wide, and east of Millstone Road it is more than two hundred ninety feet high, made up mainly of a thick outwash section capped by a thin layer of till. North of Bridgehampton, a prominent interlobate angle probably represents the junction of the Connecticut and Connecticut-Rhode Island lobes of the Laurentide glacier. The axis of the interlobate zone is breached by the Long Pond Channel and pater noster lakes. The alignment of the Long Pond and Scuttlehole kettles documents the extension of glacial ice into the northwest to southeast drainage channels. The Long Pond channel cuts through the moraine demonstrating continued meltwater activity during recession of the ice. To the east, the Amagansett Moraine segment of the terminal moraine strikes southeasterly toward the coast where it is truncated by erosion at the escarpment north of Beach Hampton. Relict boulders on the sea bed south of the Montauk Peninsula show that the terminal moraine was deposited a few miles south of the present coast (Figure 5). The Amagansett segment of the terminal moraine may have been truncated by a meltwater channel carrying postglacial lake water through the Napeague Beach gap. A sinuous curve of this channel is seen cut into the cliff on the west side of Napeague Harbor. The Napeague Beach gap in the recessional moraine was subsequently filled by the longshore current with barrier beach sands. Between the west and east arms of the moraine, south of the interlobate angle, the outwash plain opens southward and is cut north to south by a network of branching drainage or meltwater channels. Today, the southern margins of these channels end in tidal estuaries and ponds sealed off from the open sea by barrier beaches.

## RECESSIONAL MORAINES

When the last glacier began to recede shortly after twenty two thousand years ago, recessional drift was deposited north of the terminal moraine. Where the dynamics of glacial flow caused the ice to pause during the wasting process, recessional moraines were deposited. These stands of the glacier are marked by linear trends of kames. In the South Fork, the main recessional position of the Laurentide glacier north of the terminal moraine is the Sebonac Neck-North Haven-Prospect Hill Recessional Moraine (Figure 12). To the east, this moraine correlates with the Beacon Hill Moraine on Block Island; its westerly extension has not yet been traced beyond Sebonac Neck. The moraine includes kames in Sebonac Neck, North Sea, Jessup Neck, North Haven (the interlobate position), Northwest Landing, and Hedges Bank, and deformed beds at Cow Neck, all of which overlie the lower drift. The "Retreat House" kame in North Haven is about eighty feet high, while the kame at the north end of Barcelona Neck rises to thirty feet. This ice stand can be visualized by placing the glacier along the north shore of the South Fork with the necks and the kames adjacent to the ice. At Sebonac Neck, the kame is covered by more than three feet of till, while at Prospect Hill the till unit is nearly ten feet thick. Generally, the tills of the recessional moraines are discontinuous and can range from several feet of sandy till to a single layer of till stones that in this region display Connecticut and Rhode Island provenance as well as northerly fabrics.

The recessional moraine is interspersed with bogs, ponds and drainage channels. The proximity of the lower drift to the surface shows that it is the confining layer for surface water bodies. Ice-shoved and ice-contact kamic deposits- one of these borders the

Figure 12. Terminal Moraine (TM), Recessional Moraines (RM), Outwash Plains (OW), Meltwater Channels (MW) in eastern Long Island.

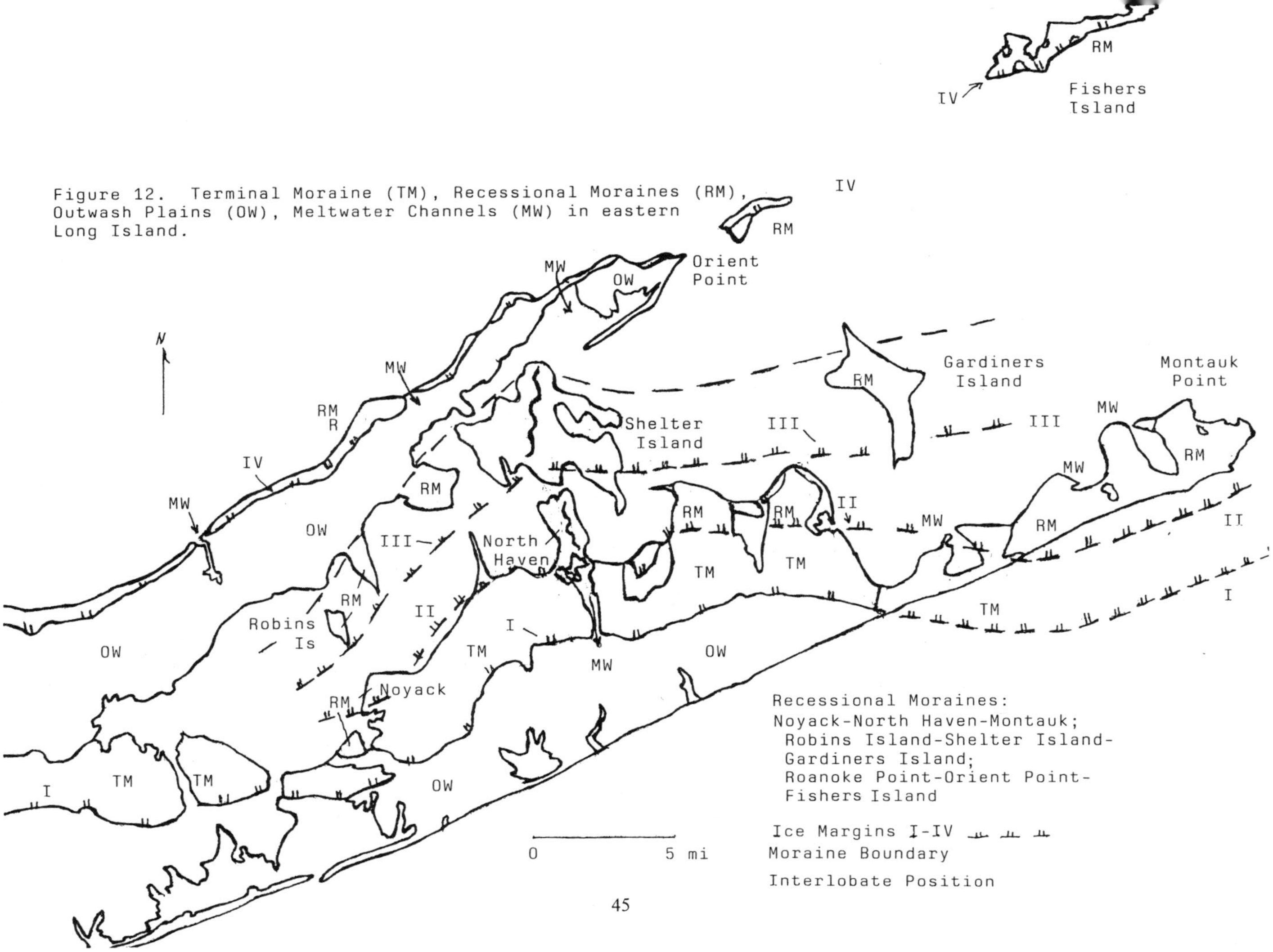

east side of the old Montauk landfill- make up part of the Prospect Hill Recessional Moraine where the upper drift drapes over the lower drift surface. The recessional moraine in the Hither Hills area has a washboard, ridge and run-off channel topography formed by a glacier that receded in a short-term rhythmic, stop and go fashion. The recessional deposits thicken across Northwest Woods and incorporate large kames at Acabonack Cliff, Fireplace Lodge and Hedges Bank. Minor east-west channel orientation at Alewife Creek, Hands Creek, and Acabonack Harbor may be relict drainage marginal to the ice. There are also relict channels that cut into the moraine, generally from northwest to southeast, and now form bays and lakes. Prominent among these are Lake Montauk, Fort Pond Bay and Fort Pond, Napeague Harbor and Three Mile Harbor in the South Fork, and Hashamomuck Pond, Maratooka Lake and Mattituck Harbor, and Port Jefferson Harbor on the North Fork and North Shore. Napeague Harbor and Fort Pond in the Montauk Peninsula and Great Salt Pond in Block Island appear to have breached the moraine and carved graded channels toward the late Wisconsinan sea. Meltwater from this glacial stand also cut deep channels into the drift to the south in the South Fork, and meltwater channels can be traced between Cold Spring Pond southward to Old Fort Pond, and from Bullhead Bay to Heady Creek. A sinuous channel may have existed between Scallop Pond-West Neck Creek and Little Sebonac Creek toward Big Fresh Pond, with a tributary from North Sea Harbor, and Little Fresh Pond. From there the channel may have extended to Agwam Lake and either west to Heady Creek or east to Mill Pond and Mecox Bay.

The next stand of the wasting glacier deposited a recessional moraine north of the South Fork. This recessional moraine is aligned from west to east connecting Robins Island, Shelter

Island and Gardiners Island, all are islands located between the South and North Forks, and Little Hog Neck and Great Hog Neck, the south-pointing peninsulas of the North Fork that lie between Robins Island and Shelter Island. This lineation of morainal segments correlates with the Corn Neck Moraine of Block Island. The re-entrant angles of the interlobate zones fall in Shelter Island and Gardiners Island, the site of ice-shoved glacial and marine strata. The areas between the moraines were invaded by meltwater that was then impounded in proglacial lakes where lake sediments were deposited. A large tract of lake clays underlies the Smithtown region east of the Manetto Hills Interlobate Zone and north of the Ronkonkoma Moraine. Lake sediments are also found beneath the marine deposits of the Peconic lowland, while small scallop-shaped wetlands dot the outwash plain between the Ronkonkoma Moraine and the Roanoke Point Recessional Moraine to the north.

The last major ice stand on Long Island coincides with what is now the north shore, extending from the Sands Point Recessional Moraine of the Hudson Glacial Lobe in western Long Island to the Roanoke Point Recessional Moraine of the Connecticut Lobe in eastern Long Island. These two recessional moraines form an interlobate angle in the vicinity of Smithtown Bay. To the east another interlobate position, or angle, is inferred between East Marion and Orient; the Roanoke Point Moraine lies to the west of this angle and the Orient Point-Fishers Island Recessional Moraine segment lies to the east. Much of the eastern part of the recessional moraine forms the modern shoreline and coastal processes have essentially cut it in half-the long way-so that only the leading edge, or distal slope of the moraine, remains.

The Roanoke Point Moraine is made up of till in varying thicknesses over outwash, like most of the other moraines. The outwash section also contains the typical deformation structures and masses of thrusted sediment, including lacustrine silts and clay, as well as flow tills. Postglacial loess covers the upper till of the sequence. The moraine describes a north-facing arc from Old Field Point to Orient Point which geographically suggets the shape of the Connecticut glacial lobe. While this crescentic form might be mistaken for an erosional remnant of the moraine, till fabrics and till stone provenance studies convincingly verify the lobate origin. First, till fabric directions shift from northeasterly at Port Jefferson to north and northeast at Wading River. Then they trend north to northwest at Roanoke Point and Mattituck, and finally northwest in the cliffs at Southold. Interestingly, the fabrics vary from northwest to northeast at Orient Point possibly an indication of the interlobate position. Secondly, Connecticut Valley Mesozoic rocks are prominent in the western half of the Roanoke Point Moraine, while granites and gneisses dominate in the eastern reach.

The Fishers Island segment of this recessional moraine is a little more complicated. This island is a complex of three overlapping ice margins that forms the link between the Orient Point segment and the chain of small islands- Plum, Great Gull, and Little Gull-rocks and shoals to the west, and the Charlestown Recessional Moraine of southern Rhode Island to the east. The oldest and most southerly Fishers Island margin is a lineation of kames and intervening kettle lake basins. The lakes are now submerged embayments of Block Island Sound, but retain the arcuate shape of the original kettle shoreline, and late-glacial lake peat underlies recent marine sediment in the basins. The middle position is a bouldery moraine of coarse meltout till and proglacial lakes dammed against the deposits of the first margin. Proglacial lake deposits appear to have been ice-shoved against the proximal margin of this moraine. The third position is interpreted from the chain of rocks, shoals, and islets running parallel to and north of Fishers Island in Fishers Island Sound.

POSTGLACIAL SEDIMENTS, CHRONOLOGY, AND VEGETATION

The chronology of late Pleistocene events is keyed to radiocarbon ages of organic-rich sediments. At Port Washington in western Long Island, glacial till containing spruce wood has been dated at twenty two thousand years (Figure 10). This date establishes the time of arrival of the Hudson Lobe of the late Wisconsinan glacier and the deposition of the terminal moraine. Assuming the Connecticut and Connecticut-Rhode Island lobes of the glacier that formed much of eastern Long Island were synchronous with the Hudson Lobe, then the terminal moraine of eastern Long Island was deposited at the same time. We can also assume younger ages- perhaps as much as two thousand years- for the recessional moraines: the lineations of ice-contact deposits through Sebonac Neck, Prospect Hill and Beacon Hill on Block Island; Shelter Island, Gardiners Island and Corn Neck on Block Island; and Roankoke Point and Fishers Island.

If so, Long Island was ice free around twenty thousand years ago. This chronology is confirmed by radiocarbon dates indicating that nineteen thousand years ago the Hudson Lobe of the Laurentide glacier had receded to the latitude of southern Westchester County, New York, and northern New Jersey, both well north of the terminal moraine. With this timetable, it is apparent that the last ice sheet did its work, left its deposits, and departed in relatively quick order.

As the glacier retreated from eastern Long Island, it stood long enough where the north coast is today to deposit a major recessional moraine, the Roanoke Point Moraine of the Connecticut Lobe. This moraine is aligned along the trend of hills and islands that link it with the Sands Point Moraine of the Hudson Lobe in western Long Island, and to the east to the Fishers Island Moraine and the Charlestown Moraine of the Connecticut-Rhode Island sublobe (Figure 5).

To understand what happened on eastern Long Island after the ice left twenty thousand years ago, we can examine some of the established climate and vegetation responses. As continental glaciers advanced, cooler climate forced vegetation to migrate southward into more favorable regions. It is not that plants can pick up their roots and run for it, but successive generations propogate where soil and climate are more suitable. In eastern North America, the forests and ground-cover vegetation had already achieved both latitudinal and altitudinal preferences over the course of millions of years of climatic change.

Shifting patterns of vegetation occurred both before and after the mid-Wisconsinan warm interval around thirty thousand years ago, as well as during and after the last glaciation. During the mid-Wisconsinan warm interval, oak forests, indicators of temperate climate, succeeded the earlier spruce forests that had been established in a much cooler climatic regime. The oaks were later replaced by forests of spruce and pine indicative of climatic cooling. During full glaciation around twenty-two thousand years ago, oak-hardwood forests had migrated southward as far as southern Florida. Spruce and pine, responding to cooling climate ranged into northern Florida, and a swath of Arctic tundra filled in along the ice front. In some areas, climatic refuges enabled plant and animal species to survive the ice age. So while Long Island-to-be was under the ice, twenty-two thousand years ago, lands to the south, including the present continental shelf, were above sea level and vegetated by plants of varying resistance to the cold (Figure 6).

During warmer climatic intervals between glaciations, such as the Sangamon Interglacial, the mid-Wisconsinan interstadial, and the current postglacial episode, sea level rose as glaciers melted, and vegetation migrated northward. Near the receding ice front, Arctic shrubs, like birch, alder, willow, and tundra herbs pioneered on fresh outwash, and spruce thrived on rocky glacial soils and in wetlands. Later, pines became established on well-drained sandy soils. Forests of oak and hickory and other temperate species followed the pines into this region, taking over the till-covered terrain.

The postglacial record for eastern Long Island began shortly after the ice receded twenty thousand years ago, at a time when the climate was still rigorous. Winds crossing the barren land surface adjacent to the ice swept silt out of the glacial deposits and laid a blanket of the finely ground sediment, known as loess, over the moraines and outwash. Up to three feet of loess can be found covering the glacial deposits of the eastern Long Island. At the same time, the daily freeze and thaw of the wet soils enabled them to flow over the permanently frozen soils below. Excavations into the surface deposits often reveal solifluction (soil flow) bands and contorted layers resulting from this process.

Vegetation invaded this barren, glaciated landscape, and the plants shed their pollen and spores into air and waterways. These reproductive cells ultimately became incorporated into the layers of sediment in ponds and bogs. Analysis of the accumulated plant debris has provided the text of Long Island's vegetational history. This story has been reconstructed from the fossil pollen found in cores taken from the wetlands. Each layer of sediment, from the glacial deposits beneath the sediments of a kettle lake to the peat forming at its surface, yields an assemblage of pollen that represents the plants growing in the area at the time that layer was deposited. Layer by layer the pollen spectra reveal differences in climate and vegetation. In the accompanying diagram, the pollen data from four sites is summarized (Figure 13). These sites include Lake Ronkonkoma marsh, a peat-filled kettle embedded in the Ronkonkoma Moraine; North Cattail Bog, a kettle bog in the Harbor Hill Moraine just west of the Manetto Hills Interlobate Zone; Cranberry Pond, a kettle lake in the Ronkonkoma Moraine in Moriches; and Sunken Lake, a kettle lake in the Roanoke Point Recessional Moraine. In the diagram, when read horizontally, the percentage bars indicate the relative abundance of each plant at that level or time; the variety of plants suggests the nature of the overall vegetation. The vertical profiles show how the relative abundance of each plant type changes over time. The pollen record documents the northward migration of the vegetation as the ice retreated. The oldest of the late-glacial pollen-bearing sediments are found in the deepest lake beds overlying glacial sediments and below the peat bog layers. They contain tundra shrubs and herbs, including birch, willow, grass and sedge. These plants confirm the presence of tundra even as the ice receded. Pollen assemblages from this interval are collectively referred to the Herb, or T, Pollen Zone. In succeeding layers of increasingly organic sediment, pollen assemblages are dominated by spruce and record the northward advance of spruce forests. This interval is designated the Spruce, or A, Pollen Zone. The prominance of spruce forests can be traced from around nineteen thousand years to about eleven thousand years ago, about the same time that people were finding their way from Siberia to North America across the Bering Land Bridge. Similar communities of plants and plant successions, including tundra and spruce forests, are found in southern Alaska where glaciers are receding today.

The next forests are represented in the peaty sediments by an abundance of pine pollen,

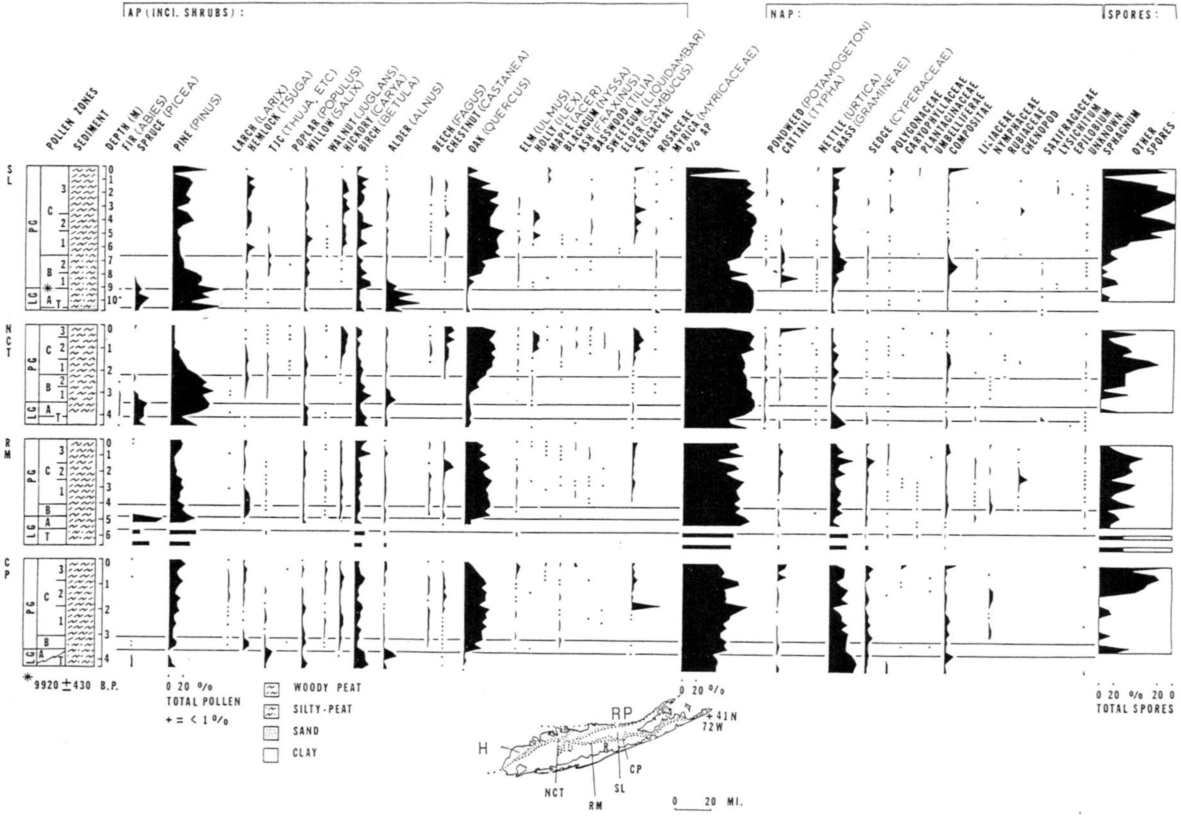


Figure 13. Pollen Diagrams for Central Long Island. Bog sites are keyed to the text. Moraines shown are H-Harbor Hill, R-Ronkonkoma, RP-Roanoke Point

first in association with birch and grass, and later with oak. Pine forest succeeded spruce in the northward migrations and prevailed until around nine thousand years ago. Although the Pine, or B, Pollen Zone is of shorter duration, its signature is consistent in all of the pollen records in this region where somewhat warmer and drier climate proved favorable to pine. Pine forests migrated northward selecting well-drained sandy tracts of outwash, lake beds and river terraces, while spruce moved further north and into higher terrain.

The onset of cool and wet conditions around nine thousand years ago enabled hardwood forests to overtake the conifers. The Oak, or C, Pollen Zone, based on the pollen assemblages in the peat that contains the upper part of the pollen record, begins with an abundance of oak and pine, then a variety of trees and shrubs, including holly, cedar, alder, birch, willow, beech, chestnut, black gum, maple, ash, heath and rose, become established. At the time of European colonization, eastern Long Island supported a complex hardwood forest on the moraines and vast grasslands and pine barrens across the outwash plain. The forest was rapidly depleted due to land clearing for farming and wood cutting for fuel. The land clearing process is documented in the pollen record in the abrupt increase in the composites, a large group of common herbs that include aster, ragweed, ragged sailor and goldenrod. These herbs invade open or disturbed ground as "weeds" competing with the crops. The composite marker stands out in the pollen profiles at each of the sites and confirms that the pollen record is intact through the colonial period (Figure 13).

It is our good fortune that the bogs have preserved the vegetational and climatic record of the postglacial interval. This record is also important because it shows that the moraines on the surface of the Island were all deposited during the same postglacial interval. If the

Ronkonkoma Moraine was the product of an earlier glaciation, the pollen records at Ronkonkoma Marsh and Cranberry Bog would show a multiple cold-warm-cold-warm pollen record. They would contain evidence of the vegetational fluctuations associated with climates through two glaciations and two warm intervals compared with the simpler cold-warm record of Sunken Lake. As it stands, the records of all of the sites are quite similar and contemporaneous.

At the same time the vegetation was changing, the land-sea boundaries of eastern Long Island and the surrounding region were being profoundly altered. The sea gradually covered the continental shelf, and the higher hills of the moraines to the east became peninsulas and islands that were then exposed to the ravages of time and geological processes. From twenty thousand and twelve thousand years ago, glacial lakes, dammed between the glacier and the moraines filled the Long Island Sound and Block Island Sound basins. Thereafter, the terminal moraine was breached and the lakes drained into the ocean. The lake waters cut through the moraines and the rising ocean gradually submerged tracts of bouldery moraine. Today, subsea segments of the moraines have been mapped between the islands as reminders of the once continuous nature of the moraines.

At that time, drainage through the Peconic area formed a channel along Gardiners Island and southward through Napeague Beach. The Montauk Peninsula was cut off from the rest of the South Fork, and the offshore segment of the Amagansett Moraine was heavily eroded. The sea continued to rise for the next six thousand years, nearly reaching its modern level and eroding a sea cliff into the Amagansett Moraine. Coastal currents gradually brought in waves of sand that filled the breach by forming the Napeague Beach tombolo, an isthmus reuniting

Montauk to the South Fork. Today the sea continues to rise at an average rate of one foot per century, and the Montauk peninsula continues to supply sand for the south shore beaches. The geologic future of eastern Long Island depends on the rate of sea level rise over the next few hundred years. Theoretically, the sea could again submerge the low lying beach and dune areas and again isolate segments of the moraines.

Recent deposits of this region include layers of peat that have formed in bogs and salt marshes, as well as the sediments of stream, marine or wind origin. Marine sands have filled old stream channels and cut some streams off from their ocean outlets, forming coastal lakes. Barrier beaches have formed along the coasts and small sand spits have extended across many of the bays. Following the emergence of the sand bars, wind-blown, fine sand accumulated in the form of sand dunes. Along Napeague Beach there is the modern primary dune line parallel to the beach, as well as older dune lines that swing more to the northeast toward Promised Land. The most impressive of the eolian features are the "walking dunes" of Hither Hills, a set of dunes that are actively migrating toward the southeast and burying the oak and pine forest in their path.

## GEOLOGIC PROCESSES

Since the melting of the last glacier, the ice-free surface of the moraines region has been exposed to the atmosphere. Eastern Long Island includes glacial and coastal terrain that has stood against the weather for over twenty thousand years. Keep in mind that after glaciation the rising sea did not reach the terminal moraine for nearly eight thousand years. During that interval, the sea gradually encroached on the continental shelf-coastal plain. Eastern Long Island was part of a long ridge with a few high hills, many of which became the offshore islands. As sea level rose to the moraine, the region probably looked something like it does today- a straight coastline of barrier beaches, headlands and bays- only much broader and with long arms of the sea between the terminal and recessional moraines.

### WEATHERING

Rock fragments begin to break down after prolonged exposure to changing atmospheric conditions. Weathering involves both chemical decomposition of the mineral compounds and their physical disintegration into smaller particles. Both types of weathering are augmented by the interaction of water, which acts as a solvent in chemical reactions, eliminating certain bonding elements, and as a mechanical lubricant and physical wedge in separating mineral grains and opening planes of weakness. This activity is accelerated when water in open spaces freezes and expands.

Rain can combine with some substances to form weak acids and bases that further weaken chemical bonds in minerals and release by-products such as water and gases. Some minerals lose their cohesiveness and eventually turn to clay; others simply break down into small particles of sand. Feldspars and micas, and various iron and magnesium-bearing minerals break down rapidly in humid climates, while quartz is resistant to chemical action. Water can also be incorporated into the structure of some minerals to form a hydrous variety. Oxygen from the atmosphere combines with iron and aluminum to form oxides which can be carried into other environments.

Physically, water seeps into the smallest partings in rock, including pore spaces between mineral grains, where it can precipitate crystalline compounds or freeze and expand to create larger openings and force the minerals apart. On the surface water can saturate a thin rind of rock forming a spherical layer around the rock that will eventually peel off like the skin of an orange. We can find erratics on the beach that exhibit these weathering rinds, as well as rocks that have split from the repeated action of freeze and thaw. The more massive crystalline rocks seem to disintegrate more readily; finer-grained metamorphic rocks, like schist which has fine foliations along which water can penetrate, are more apt to decompose. It is not unusual to find boulders of schist that are thoroughly decayed.

The rate at which weathering processes act varies with the climate. We can get an idea of short-term weathering in a humid climate by studying headstones of various rock types in a cemetery. We can tell how long a certain type of rock has been exposed to the elements if the dates on the stone are intact. Comparing the condition of the rock- loose mineral grains, weathering rinds, and separation of bedding and foliation planes- with the length of time it has been exposed, provides a relative measure of the weathering rate. Some headstones will last longer because water can not penetrate the surface, while others are open to water infiltration and will rapidly decay. Slates separate along foliations over relatively short periods of time, losing their inscriptions along the way. Granites are more often affected on a grain to grain basis by freeze and thaw, expansion and contraction, and the tendency of some minerals, like feldspar, to decay. Porous sandstone markers allow water infiltration and disintegrate rapidly, and so are rarely used today. Tightly compressed and cemented sandstones and flagstones can resist physical weathering and may endure as markers. Serpentine-rich marble headstones are susceptible to air pollution and acid rain and are not widely used in an unprotected state. Polishing and coating the stones protects the decorative inscriptions, as well as the body of the rock, making the marker more durable.

## SOILS

With the melting of the ice, the glacial deposits were exposed to the dynamics of the atmosphere; wetting and drying, freezing and thawing, dissolving and acidifying in the process. Over thousands of years, the minerals in the ground-up rock debris slowly decayed and disintegrated to provide the basis of soil development, the sediment and an assortment of compounds capable of supporting plant and animal life. Rock-forming minerals like feldspar and mica, rich in potassium and aluminum, released these elements as they were converted to clays. Carbon compounds and water were given off to the soil, and iron and aluminum combined with oxygen to form oxides. Rainwater percolated into the soil dissolving some solids and redepositing them deeper in the sediments. Organic debris from plants were added at the surface along with the by-products of decay, enriching the soil and sustaining plant and animal nourishment.

Typically soils in humid regions, such as Long Island, develop three distinct soil horizons or zones (Figure 14). The lowest horizon, designated the C soil zone, is most like the original or parent material: bedrock or transported rock debris, like glacial deposits, which range in particle size from clay to boulders. Above the C zone is the B soil zone, which consists of sediments that have been considerably altered from their C zone source. This zone often has a high water content due partly to the water table that can collect on an impervious C zone, such as a layer of glacial till. Soil water seeps through the B zone carrying compounds of clay, iron and aluminum oxides, carbonates and humic acid. These materials are redeposited lower in the B zone, which is appropriately termed the zone of accumulation, where they form bands of mineral enrichment like those cemented by red and yellow-colored, iron and aluminum oxides.

The uppermost soil zone, the A zone, includes an upper,organic-rich and life-sustaining layer with abundant roots and organic litter at the surface, and typically, a lower white-colored and leached horizon that supplied oxides to the B zone; thus the name, zone of leaching, is applied to the A zone. Plants supply the organic compounds, and rainwater mixing with humus and carbonates forms weak acids that help to dissolve A zone minerals. In the last century, rainwater has become more acid due to sulfuric aerosols from the burning of fossil fuels, thus adding to the soil acidification.

Soils formed on glacial drift are commonly called loam, a mixture of sand, silt and clay, or clay loam or sandy loam depending on the dominant sedimentary grain size. Eastern Long

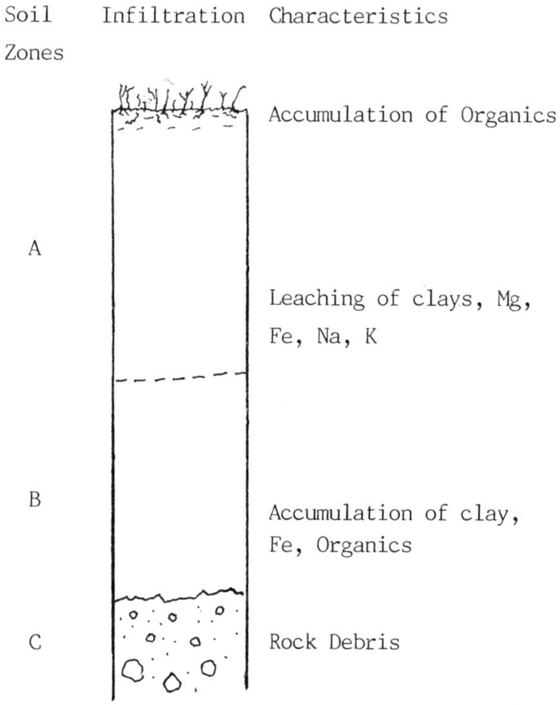


Figure 14. A Typical Soil Profile

Island loams have traditionally been given local geographic names, such as the Montauk sandy loam and the Riverhead loam, the latter a coarse textured variety of the soil derived from glacial till. The thick silt or loess cover over eastern Long Island gives rise to a silty loam, the Bridgehampton loam. This soil is the basis of a productive agricultural stratum because of the abundance of very fine-grained and unaltered, mineral particles. The sand in the sandy loams is derived mainly from quartz-rich granite and gneiss gravels. Quartz is a mineral that weathers very slowly, breaking down into smaller grains rather than decomposing, and because of its resistance, it is the primary ingredient of beach sand.

Technically, these Long Island soils are pedalfers, those rich in aluminum (Al) and iron (Fe), and are soils characteristic of humid and temperate climate. The pedalfers of southern New York and New England are also called gray-brown podzolic soils, and in a more recent classification, spodosols. These soils normally are initiated in a cool and humid climate in a forested region that provides abundant organic materials, much like the postglacial Long Island region. Both the loess and the glacial drift, now covered by sand dunes, may have old, buried soil profiles, termed paleosols, which are significant indicators of past soil-forming environments.

The organic soils of peat bogs are quite different in composition. They are rich in partially decomposed plant debris but have little of the mineral content of the loams. Pollen, spores, seeds, needles, and other tough plant particles are preserved in the layers of peat. Many glacial lakes have been filled with lake bottom sediment and then covered with peat. While lakes support aquatic vegetation, increased nutrients in the water, including terrestrial pollutants, can generate population explosions, or blooms, of algae and grasses that can threaten a lake's existence. Left uncorrected, this process leads to rapid filling of the lake with organic debris, a common fate of many glacial lakes. These layers of organic sediment have provided the record of vegetation and climate in the variety of pollen that has accumulated since the last glaciation. As discussed earlier, this important aspect of geologic history can only be found in these lake and bog sediments.

Sand dunes and beach deposits normally have very little soil development and very weak soil zones. These soils, known as entisols, may support plants, such as dune grass and beach rose, that are adapted to this environment. Soil development on these sediments can also be interrupted by coastal erosion due to storms.

## MASS WASTING

Mass wasting involves the movement of rock materials downslope from where they are weathered, in response to the pull of gravity and with the aid of water. The process ranges in magnitude and rate of movement from the nearly imperceptible creep of soils on gentle slopes to the abrupt fall of rock debris from cliffs and the slumping and sliding of massive blocks of rock and sediment. The slumping of masses of morainal deposits from the headlands along the coasts is a familiar example in eastern Long Island. Movement can be of sufficient magnitude to damage or dislodge surface features, including buildings, whether cottages or lighthouses, perched on the edge of a sea cliff.

Falling debris builds cone-shaped piles of sediment at the base of cliffs, while individual boulders just tumble down to the beach. Large erratics lodged in the moraine overhead are the local equivalent of a falling rock zone as they sit precariously with only a little sediment to hold them back. Downslope movement of sediment can be initiated by water which saturates fine-grained materials, turning them into liquid mud. Beds of clay eventually become saturated with water and can form "quick clays" that flow when a short-term disturbance- any kind of shock, ranging from an earthquake to a footfall-triggers the flow.

A major type of mass wasting along glacial coasts is slumping. This process is the result of a number of factors that include rainwater infiltration into the sediments, saturation of clays, erosion of the cliff face, wave erosion at the base of the cliff, and the orientation of the layers of sediment and rock. Slumping is a prominent feature of eastern Long Island bluffs where coarse and compacted sediment, such as glacial till, often slides downhill as a cohesive mass. Water-lubricated layers within the sediment and wave erosion, which cuts away the base of the bluff, aid in the process.

Slumps begin when ground water seeps into porous soil and migrates downward, seeking out planes of weakness. These planes form around dislocated blocks of lower drift and along surfaces of impermeable sediment. They gradually evolve into fractures, or faults, that extend downward from the seepage zone to the base of the slope, taking the form of curved planes of rotation (Figure 15). We might suspect that one of these fractures is beginning to form when a line of robust plants emerges along the surface where rain water collects parallel to the edge of the cliff. The plants are responding to the water that seeps into the initial opening of the fracture. As the slump progresses, a small offset or sag, like a miniature cliff, will appear. When the space between the slump block and the bluff becomes large enough, water will

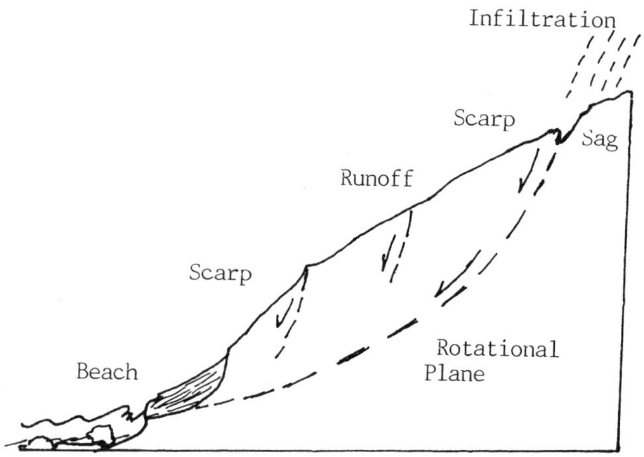


Figure 15. Profile of a Sea Cliff Slump Block

accumulate to form a sag pond.

Water entering the rotational plane continues to lubricate this fault zone and encourage slippage. Irregularly-shaped, sliding blocks of the lower till are gradually etched out by rainwater erosion and whittled into deep gullies, monuments and pinnacles, forming a jumbled badland of odd-shaped remnants, known as hoodoos, between the bluff and the beach. Old slumps can be recognized because their block-shaped remnants stand out from the wall of the bluff. The step-like tops of the slumped blocks are still covered with growing vegetation, but they are lower than the original surface and are usually tilted back toward the upland, with their transported plants bending toward the light. These outliers are actually slowly rotating slumps detached from the bluff and sliding along the curved fault plane. With on-going wave erosion at the base of the bluff, the block migrates downward toward the beach and its eventual destruction in the surf. As more water is added to this system, rapid movement can be triggered. When fine-grained sediments make up some of the layers, mud and debris flows may accompany the slump. While slumps, slides and flows are common types of slope failure on coastal bluffs, they seem to develop without warning. The slumping process may be slow, with all of the characteristics of mass wasting present, and slopes are in a continual state of adjustment. We have also learned that natural processes are accelerated by construction projects. Excavations, such as cellar holes, water and drainage pipelines, and septic systems, disrupt the sediment layers and make it easy for water to enter the soil. Any activity that reorganizes sediment layers will enable water to infiltrate the surface. Any channel dug into the sediments will become a conduit for water, even when refilled with the original sediments. This is especially true when pipelines are installed in the excavation.

Recently when the Southeast Lighthouse on Block Island, Rhode Island, was moved, a large amount of sediment was excavated from around the building and the old cellar. After the move, soil was replaced and the surface graded. However, the reorganized glacial sediments no longer had the impermeability inherent in the original layers of compact glacial till. Now surface water can more readily infiltrate the porous sediments of the headland. As a consequence, saturation of the sediments by rainfall may accelerate the rate and the direction of erosion into the bluffs and into the original site of the lighthouse. The Montauk Point Lighthouse, which is surrounded by several slump blocks, could also be jeopardized because of a long history of excavations- for example, wells, septic systems, and cisterns- around this site. However, a long-term maintenance and erosion control project has armored the Point with several tiers of large blocks of crystalline rock. Although dewatering of the surface has not been attempted, water runoff is controlled, in part, by the abundant grasses and shrubs, planted in step-like terraces. Constant repair of the sea wall and step complex should extend the life of this precarious promontory to the extent that coastal erosion might someday create a separate small island with a lighthouse surrounded by a rock wall.

Slumps and eroded blocks of sediment are highly visible along the coasts where bluffs have undergone significant erosion during recent storms. Major storms have increased the rate of mass wasting well above the long-term average of one to three feet per year, depending on the exposure. In many headlands, sand layers have been washed out leaving large cavities around masses of lower till. Due to lack of support, the remaining, more resistant till will eventually slump, and the bluff will recede. Overhangs formed where roots hold the soil layer, fresh slump blocks of till, rock falls, deepened gullies, and sediment fans are all visible components of slope readjustment, a process that continually tries to bring the eroded surface of the land into balance. Because the more gradual recession of eroded slopes is not as dramatic as a storm's rampage, we are not aware of the full extent of the damage until we are faced with the devastating loss of property.

## BEACHES

Most beach walkers today keep a close eye on the seasonal changes that affect the narrow sand bars along the eastern Long Island coast. Those who watch the shore during and after storms also see how the waves vary in height, turbulence and direction as the storm passes. We are well aware that beaches change shape because of erosion and deposition, and we see the short-term effects of the tides. In recent years, the surf at high tide has often invaded the back beach. In the big storms of the early 1990's, high tides and storm surges cut cliffs in the dunes, excavated bluffs, and washed millions of cubic yards of sand from the beaches. The catalogue of major changes along the coast must include barrier beaches where the sea has washed away large quantities of sand, roads, and beach houses and cut new inlets.

In a general sequence of beach development, beaches are annually stripped of sand by high energy waves generated by winter storms. They are replenished in summer when gentler conditions prevail and low energy waves push layer upon layer of sand back onto the beach. On most beaches, sand covers a wide expanse of land between the ocean and the dunes during the summer. So much so that it is hard to imagine the dramatic thinning of the sand layer and narrowing of the beach that takes place in the winter. Where headlands of glacial moraine abut the coast, wave erosion exposes ridges of coarse lag gravel cut into neat scallops by the waves. The process of forming these beach cusps results from the complex geometry of wave direction and size, the nature and abundance of the materials, and the orientation of sand ridges on the offshore sea bed.

As summer approaches, moderating weather reduces the wind-driven wave energy, and the sand begins to replenish the beach. Slowly the gravel bank vanishes beneath the sand sheet, and the beach grows wider and the sand layer thicker. Storm-driven tides of early spring that swirled into the dunes are replaced by a more regular and less destructive rise and fall of the sea. Only after a passing storm do we find the physical evidence that a new layer of sand has been spread across the beach. Wave erosion has carved terraces into the sand, and surface runoff from the beach cuts gullies through the sand layer into the underlying gravels. This sequence of deposition and erosion is repeated for each storm, and between storms, the sand layer may be gradually rebuilt.

Where, we might ask, does the sand go in the winter? And where does it come from in the summer? Is the sand stored offshore and recycled back and forth by the season? Or perhaps this is all new sand brought down from the bluffs and pushed along the coast by wind and waves? The answers to these questions are not entirely reassuring. For example, sand may be stored in offshore beds, but much of it is permanently lost from the coastal zone, carried out to sea by strong, storm-generated bottom currents, and into water too deep to be reclaimed by summer waves. Some sand settles in harbors and can only be retrieved by dredging, while the fine-grained sediment on the beach accretes to the dunes through wind action. In fact, most of the sand for each new summer season comes from the erosion of the headlands and beaches that lie upstream along the coast. The greater the erosion of the bluffs, the more the beaches are nourished and the healthier they seem by mid-summer. Recent coastal storms have also cannibalized beaches along the Atlantic coast, sweeping sand into the bays and inlets, piling it up against jetties, and washing it out to sea.

To better appreciate the imbalance between sand supply and beach formation, we have to examine the origin and development of beaches, their morphology and their recent history. Typically, the beach is a complex of landforms made of deposits of sediment of varying sizes, sorted out by the differing energy levels of waves, wind and tides. It extends from the edge of the ocean across the beach and dunes to the bayside (Figure 16). The forebeach reaches from the low tide line landward through the berm, a ridge above the high tide zone that runs parallel to the shoreline. Sand accretes to the forebeach and expands the beach face seaward. At high water, waves spread sand over the forebeach, but during storms, vast amounts of sand can be carried across the beach and into the bay as washovers. Offshore, deep waves and currents reshape the sand sheet on the sea bed into submarine ridges with intervening swales or runnels, demonstrating the energy levels that reach into deeper water. Strong waves form impressive breakers, and because of their height, they also have an impact on the seafloor. There can be little wonder how storm waves remove sand from the coastal zone and generate counter currents that move the sand seaward.

Landward, the back beach consists of a low plain between the berm and the primary line of sand dunes. A depression in this plain may retain a shallow pond of sea water washed over the berm. Outlets from these ponds are rare, so the water slowly evaporates. As the water level of the pond drops, the warm surface in the swale may become encrusted with a mat of

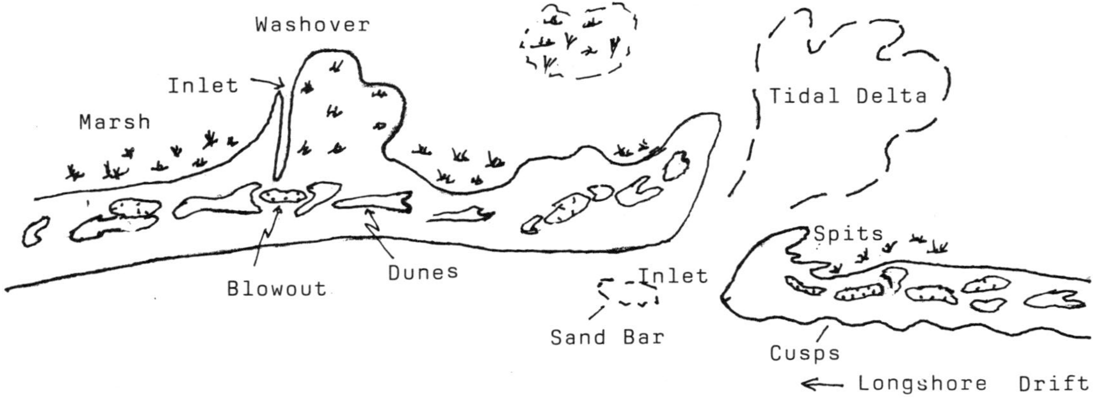


Figure 16a. Map of a Barrier Beach and Bay

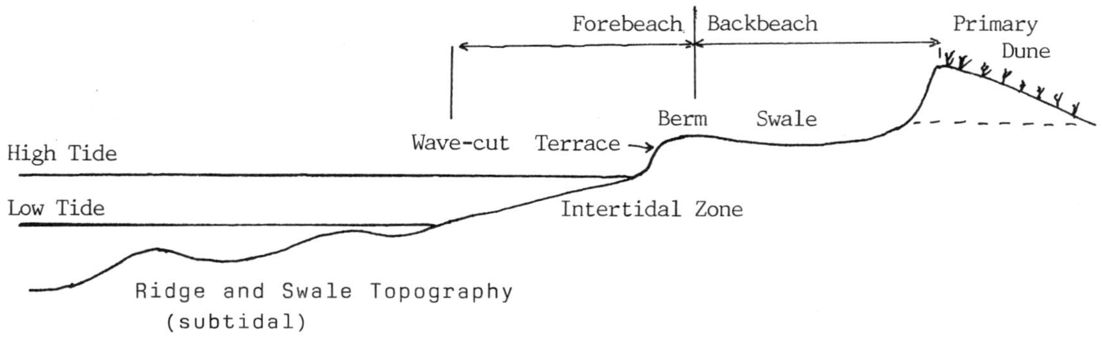


Figure 16b. Cross Section of a Barrier Beach

a) Air view of Barrier Beach- Ocean (upper left) to Bay, with dune lines and interdune swale.

b) Forebeach with Spit, Primary Dune and Interdune Swale.

Figure 16c. Southeastern Long Island Beaches

c) Eroded dune with cross bedding.

d) Saltmarsh on bayside with drainage ditch.

Figure 16c. Southeastern Long Island Beaches - continued.

algae. With higher sea levels, the back beach may be inundated more frequently, and high tide washovers can reach into the dunes. Beyond the primary dune line, older dune ridges may have formed close to the bayside. In places where washovers have cut across the beach, fans of sediment extend into the bay as subaqueous deltas. These fans support the growth of aquatic grasses, which in turn trap more sediment and build small islands in the bay. Powerful storms can also cut deep channels through the beach, thus creating new inlets.

Along the beach front, the energy of breaking waves forms a zone of coastal erosion, sediment transport and deposition parallel to the beach. Because the waves usually attack the beach at an angle, a current is generated that transmits wave energy along the beach, pushed by the waves. This current, the longshore current, transports sand down beach in the process called longshore or littoral drift (Figure 17). In this way, sand from the Montauk bluffs reaches Napeague Beach on littoral currents from waves driven by easterly winds. The source area of these winds, the fetch, results from the circulation of winds around atmospheric pressure systems crossing the region generally from west to east. The longshore current sorts out the sand and nourishes the beach along the way. On the bay side of the eastern Long Island barrier beaches and in the Peconic bays between the Forks, the longshore currents, tidal currents, and stream-derived currents carry sand eroded from bluffs along the coast to form spits that extend downstream. The prevailing direction of the longshore currents varies from headland to headland, responding ultimately to the dominant, wind-driven current or tidal direction.

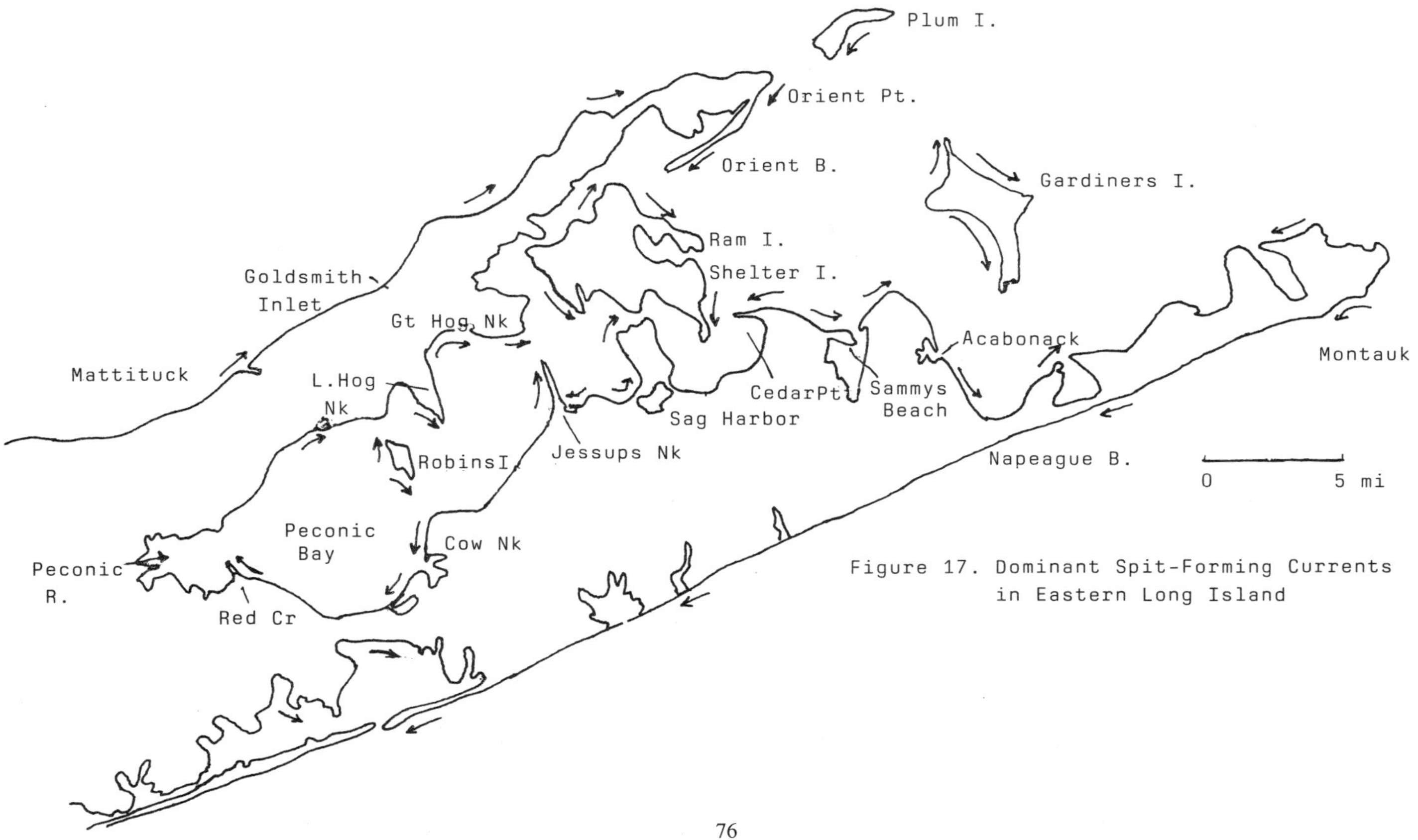


Figure 17. Dominant Spit-Forming Currents in Eastern Long Island

BARRIER BEACHES

The large supply of sand along the south shore of eastern Long Island and a relatively stable shoreline have maintained a chain of elongated beaches and dunes for thousands of years. Today, these narrow islands form a unique trend of overlapping barrier beaches from Napeague Beach to New York Harbor. Thus, the south coast of eastern Long Island is characterized by eroding headlands to the east and barrier beaches enclosing large bays to the west. The barrier beaches rise above the high tide level and extend parallel to the coast, following the prevailing westerly direction of longshore drift. They have typical beach morphology including all the topographic components of the forebeach and backbeach, along with primary and secondary dune lines, washover sand bodies, and old inlet scars.

Geologists argue that barrier beaches can evolve from emergent offshore ridge and runnel topography or from spits extended from headlands by longshore currents. The glacial headlands of eastern Long Island have been ravaged by the ocean for over twelve thousand years, ever since postglacial seas rose to intersect the moraines. Sand eroded from this upland continues today to move westward. So it would seem that the barrier beaches of southern Long Island originated from a spit that initially grew westward from the first outpost of the terminal moraine- the offshore segment of the Amagansett Moraine- to be inundated by the sea. As the terminal moraine was eroded away, the barriers merged into a once more-massive Montauk Peninsula, which has been eroding ever since.

There is an alternative opinion on both the origin and the fate of barrier beaches. At the height of the last ice age, the shoreline was near the edge of the continental shelf, as much as eighty miles from its present location. This late Wisconsinan shoreline and its sand bars migrated landward as the glaciers melted and the level of the ocean rose. The remains of submerged beaches and salt marsh peat have been found out on the shelf. With no headlands nearby- the terminal moraine was well north of that shoreline-how else could the barrier beach form?

Many coastal geologists now believe that the barrier beach system originated in offshore sand bars, piled up by the late Pleistocene ocean waves around twenty-two thousand years ago. Strong longshore currents were able to rework the abundant coastal plain sands into elongate bars. The juvenile barrier beaches gradually emerged from the sea, as winds and waves moved sand over the bars. For the next ten thousand years, these beaches migrated: westward on longshore currents and northward by washovers and accretion, as the oceanside shoreline eroded and the bays filled in. Around twelve thousand years ago, the sea reached the moraines, and the migrating sand barriers and spits from the headlands merged and were replenished by the coastal erosion of the glacial moraines.

Shortly thereafter, the terminal moraine southeast of Amagansett and south of Montauk was heavily eroded and then submerged by the rising sea, leaving the bluffs of the Montauk Peninsula to confront the waves. The beaches gradually migrated landward, and today make up part of the complex of coastal landforms exposed to coastal erosion. Ultimately, the headlands will diminish, the Montauk Peninsula will be separated into small islands, the bays will narrow to tidal channels, and then these features will disappear entirely.

Montauk's morainal bluffs have been under attack throughout Holocene time, an interval of ten thousand years. The recessional moraine can be physically traced westward to Hither

Hills. Between Hither Hills and Devon there is a gap in the glacial deposits-the western end of the recessional deposits in Hither Hills appears in the railroad cut southeast of Napeague Bay. The gap was probably cut twelve thousand years ago by meltwater draining from proglacial lakes in Gardiners Bay and Long Island Sound. Northwestward- migrating sand bars sealed the breach and created Napeague Beach as they coalesced with the coast. The late Wisconsinan terminal moraine re-emerges at the cliff north of Beach Hampton in Amagansett-the point from which the moraine originally extended southeastward into the area now beneath the sea. The continuation of the moraine from Hither Hills lies to the north, bordering the north shore of the South Fork in the vicinity of Fireplace.

Field studies of the glacial deposits have shown that the Montauk Peninsula is not the terminal moraine. Rather, it is comprised of two, superimposed recessional moraines, one from each of the two glaciations to affect this region. The terminal moraine, or what is left of it-mostly well-washed, bouldery rubble- lies submerged a few miles offshore of the Montauk Peninsula. Sand eroded from the moraine and from the outwash plain south of Bridgehampton has sealed and straightened the coastline from East Hampton to Southampton, smothering streams and forming ponds like Georgica Pond and Mecox Pond. Through the rather heroic efforts of local baymen, these ponds have been opened to the sea for hundreds of years. Promised Land sits on a thick pile of sand that accumulated during the Holocene Epoch.

Since merging with the eastern peninsula to form a tombolo with the main body of the South Fork, the barrier beach continued to migrate north and west, joining with the outwash plain at Southampton, Westhampton and Mastic Beach, and isolating Shinnecock Bay, Moriches Bay, and eastern Great South Bay. The relentless longshore drift has maintained the ever-migrating, straightened shoreline of the barrier, while storms have opened and waves have closed innumerable inlets along its length. The modern inlets-permanent by virtue of being physically maintained by governmental agencies- include Shinnecock, Moriches, and Fire Island inlets. Shinnecock and Moriches inlets originated during some long ago storms and face competion from new inlets cut by recent coastal storms. A number of other older inlets have had a brief tenure and have long since been filled. The Fire Island barrier beach has created Fire Island Inlet by gradually overlapping the Oak Beach end of the Jones Beach barrier. Whether the two barriers were connected at one time and separated by storms has not yet been established.

## SAND DUNES

The expanding sand plain of the summer beach promotes another important process- the accretion of sand dunes. If you have ever tried to sunbathe while a strong wind is blowing, you know the sting of moving sand. You might also be aware of how the sand begins to pile up against you. In effect you have become a barrier, or wind shadow, to the movement of sand, a role more successfully played by inanimate objects. When the wind is blowing hard enough to carry a particle of sediment a distance of about one foot in the time interval of one second, it is capable of carrying all material ranging in size between silt and medium sand. Over a year's time, the wind may also blow from several directions. But the net effect of sand movement will be governed by the most consistent wind direction.

As the sand moves, the grains get blasted by smaller particles and bumped along by larger ones. The sand blasting frosts and rounds the grains, while winds of variable speed sort the particles by size and weight, all the while moving them downwind. Ultimately, an obstruction, such as a piece of drift wood or the spokes of a snow fence, will form a wind shadow, and the sand will begin to accumulate. As the deposit grows, it acquires a predictable form: a long, gentle slope facing the wind, and a short, steep slope facing away from the wind (Figure 18). The sand climbs the gentle windward slope and collects at the upper edge. When the mass of sediment reaches a critical angle- around thirty degrees- it cascades down the steep leeward slope forming inclined beds similar to the foreset beds of a delta. Steady winds and a good supply of sand insure the growth and maintenance of the dunes; excessive winds and a short supply of sand cause sand to erode and dunes to migrate; a balance of wind and sand will stabilize the dunes and allow vegetation to grow on the gentle windward surface. Close inspection of the dune shows that the sand actually moves along the windward surface as wind ripples- dune forms in miniature- all lined up in parallel rows by the prevailing wind. As the wind blows, the sand grains migrate down wind from ripple crests to troughs.

Layer upon layer of dune sand accumulates in this way, eventually building dunes that reach heights of tens of feet. In migrating dunes, the beds of the gentle windward slope overlap the steeper, lee slope beds. Along the Napeague, Hamptons and Fire Island barriers, well-formed dunes rise to about thirty feet above sea level. In general, dunes are found on all eastern Long Island barriers. They also cover the upland surfaces including the tops of bluffs. Since dunes are dynamic landforms, modern dunes can hide colonial structures and artifacts, such as trash heaps, stone walls, and old foundations, as well as structures and debris of recent vintage. When dunes erode or migrate, the buried treasures are eventually revealed.

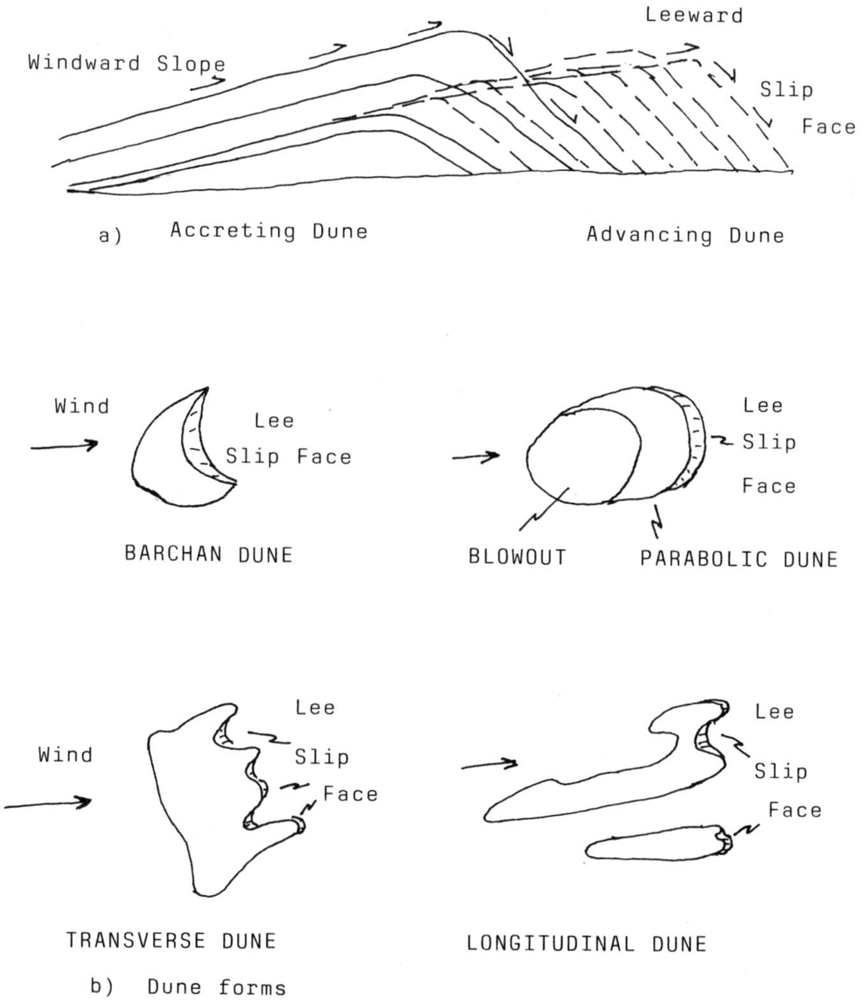


Figure 18. Profile (a) and Map Views (b) of Sand Dunes

In the erosion process usually the lighter mineral fraction, mostly quartz sand, is removed, and a lag deposit made of heavier mineral sand, including magnetite and garnet, remains as a reddish-black layer. Like quartz sand, the heavy minerals were weathered from the crystalline rocks brought to this region by the glaciers. Even these tiny particles can be specific to one source area and are important clues in the study of glacial provenance.

Another important comparison is that dune sand is finer grained than beach sand, and ocean waves are more powerful than the wind. When the sea's energy is expended against the coast, the dunes virtually melt away, reminding us how fragile and temporary they are. Thus the development and preservation of coastal dunes in this region is of considerable importance in the attempt to stabilize and maintain the shoreline. At Napeague Beach, several lineations of coastal sand dunes are derived from the eastern edge of the moraine, the source of fine sand to the dunes and coarse sand to the westerly beaches (see Field Trip 1). The dune lines extend northeasterly at slightly different angles to the coastline and are comprised of different dune forms. The type of dune is the product of wind speed and sand supply, while the alignments signify shifts in the prevailing winds over time. For example, in the Napeague Beach area, there is a line of primary dunes adjacent and parallel to the beach. A secondary trend lies parallel to the first between the back beach and Montauk Highway. Many of these dunes are elongate southwest to northeast as longitudinal dunes, aligned in the direction of the prevailing wind, and with the horn shape of barchan dunes appended to their easterly ends. Large dry and sand depleted areas, known as blowouts are interspersed among the dune lines. A line of transverse and barchan dunes curves to the north northeast south of Cranberry Hole Road, while on both the west and east sides of Napeague Harbor, much larger dunes rise above the bay. At Napeague on the west side the transverse dune rises to ten feet above sea level. In Hither Hills on the east side, the "walking dunes" are over forty feet high. Both sets of dunes are migrating to the southeast on dominant northwest winds. The Hither Hills dunes are slowly covering the forest to the southeast and banking up against the moraine, while exposing tree stumps in large blowouts in their wake (see Field Trip 1).

## THE BAYSIDE COASTLINE

The irregular bayside topography can be attributed to submergence of the recessional moraines by rising sea level and to the storm-related tides and coastal currents that create washovers and cut inlets. The inlets shape tidal sediments into subaqueous deltas on the bay floors, and washovers form lobes of sand extending from the barriers into the bays. On these sand bodies, salt marsh plants trap fine sand and build thick mats of peat on the surface. These organic layers have kept up with sea level rise, growing at an average rate of one foot every century. They provide us with a long record of regional vegetation and climate. Because they are restricted from oceanic currents, the bays have developed current patterns of their own as interpreted from the shapes of sand bars (Figure 17). The elongation of sand bars is in response to the net direction of longshore drift along the coastlines in the bays.

Rising Holocene sea level has also drowned stream valleys. The stream sediments are deposited in the channels at the point where the energy of the stream diminishes and tidal currents take over. Streams flowing into the southern bays have supplied sand for sand bars that extend, mostly as spits, southward into the bays. Circulating bay currents, driven by tides and winds, bend the pointed spits into hooks that curve either to the east or the west. Dominant northwesterly winter winds generate currents that reorient some spits, bending them to the southeast. In open larger water bodies, from Flanders Bay, Great Peconic and Little Peconic bays through Shelter Island Sound to Gardiners Bay and into Block Island Sound, complex currents have eroded glacial bluffs and sent fingers of sand like guideposts along the coast.

A tour around the eastern bays with the help of a map can trace the various currents and identify the more prominent sand bars (Figure 17). Beginning on the south side of the mouth of the Peconic River, a west to east trending spit forms Iron Point and partially encloses Reeves Bay. To the east along the south shore, small north and northwest pointing spits seal off streams to form small coves and ponds. Littoral drift to the northwest has extended Red Cedar Point into Great Peconic Bay, while bay currents have trimmed Sebonac Neck, a segment of the recessional moraine, from northeast to southwest, hung a barrier across Cold Spring Pond, and pushed sand bars across Sebonac and Little Sebonac Creeks. North of Cow Neck Point, Robins Island, a remnant of the next younger recessional moraine to the north, has sand spits on both its south and north ends, with a double bar forming the northwest point. West-moving sand on the north end of Cow Neck has completely closed off Scallop Pond and North Sea Harbor.

The long fetch of northeast-flowing currents has subdivided Little Peconic Bay by connecting Jessup Neck, also a remnant of the recessional moraine, to the mainland with a double tombolo. Tombolos have also attached North Haven to Noyack and Sag Harbor. Sag Harbor Bay with its crescentic southern shoreline sits between North Haven and Barcelona Neck. Cedar Point beach has pushed one and one half miles westward into Northwest Harbor in response to tidal currents in southwestern Gardiners Bay, but an east flowing current along Hedges Bank has extended Sammys Beach and nearly closed Three Mile Harbor. Southeast of Fireplace the net longshore drift is southward, sending a long barrier from Gerard Park across Acabonack Harbor, while a westerly-hooking bar derived from Acabonack Cliff to the south curves into the harbor. Southeasterly currents along the cliff have created the crescent beach bordering Napeague Bay, but overlapping sand spits combine to narrow the mouth of Napeague Harbor. Fort Pond Bay and Fort Pond, relicts of a glacial meltwater channel, are sealed off north and south by barrier beaches now partly obscured by urbanization. The northwest to southeast trending Lake Montauk, also a relict of glacial and meltwater erosion, is now cut off to the north by a massive east to west trending, triangular sand-spit plug.

North of Napeague Bay, currents from Block Island Sound on the east and Gardiners Bay have streamlined the recessional moraine of Gardiners Island. The long, south-trending spit at the south end is fed initially by currents on both sides of the Island. To the north, a northerly flowing Block Island Sound current sealed off the Tobaccolot Pond drainage and formed a crescent beach and Tobaccolot Bay. The triangular Bostwick bar extends to both the northwest and the south. Gardiners Bay currents on the east side of Shelter Island have formed tombolos that connect the Shelter Island headlands to smaller islands: Upper Beach extends to the southeast from Cornelius Point linking Little Ram Island to Lower Beach and Ram Island, with an additional spit south to Reef Point. Similarly, Gibson Beach is both a crescentic beach and a tombolo between erosion straightened Nichols Point and Mashomack Point. Shelter Island Sound currents flowing along the south side have hooked Shell Beach to the northeast across West Neck Harbor. On northwest currents, Crescent Beach terminates in a hook into Dering Harbor at Chequit Point.

Again starting from Indian Island at the mouth of the Peconic River and going north along the North Fork coast, crescentic bays and small sand spits are common mainly on east-moving currents. Two flow directions converge to form the west and east-facing beaches at the head of Cutchogue Harbor and the west tombolo of Little Hog Neck- another segment of recessional moraine. The northeast tombolo of Little Hog Neck shows a northerly drift toward Hog Neck Bay, as does the slowly forming southwest tombolo of the Great Hog Neck segment of the recessional moraine. An older, broader tombolo to Great Hog Neck lies beneath the main highway. Northbound Shelter Island Sound currents formed the north tombolo of Great Hog Neck and nearly closed several creeks. Hashamomuck Pond lies over another of the northwest to southeast- oriented glacial meltwater valleys. It has been sealed off on both the north and south sides by sand bars. In Greenport, Sandy Beach curls northward showing the dominant drift on the west side of the breakwater. Northeast flow forms the bars across the ponds and into Orient Harbor. North Road crosses a causeway on a northeast sand bar that seals the embayment over a remnant glacial meltwater channel. The very impressive Orient Beach-Long Beach barrier has grown over five miles westward into Gardiners Bay from Orient Point to Long Beach Point. Much of the sand and gravel originated in the erosion of Orient Point. This great sand bar cuts off bays and streams to create Long Beach Bay and Orient Harbor.

A sand bar at Orient Point is also being extended eastward toward Orient Point Lighthouse by currents along the north and south shores, linking up a chain of erratic boulders and submerged remnants of the Orient Point segment of the youngest of the recessional moraines. To the east, the moraine itself, includes numerous small islands and rocks between Plum Island and Fishers Island. Plum Island has been streamlined along the north shore from the Plum Island Lighthouse eastward, but with the greatest accretion of sand on the southwest point. Great Gull Island also has a southwest-trending spit, while a spit on Little Gull Island points northeast. On Fishers Island, a sand plain has formed from the moraine to the southwest point. The moraine is cliffed along Isabella Beach and the sand bar seals off several ponds from

from southwest to northeast. Sand bars on both sides of Barleyfield Cove document the convergence of west and east drift. Abundant bouldery debris along the shoreline reveals erosion of the moraine, and tidal currents have swept deposits eastward on the north side.

Westward on the Long Island Sound coast of the North Fork, currents and longshore drift are generally to the northeast, depositing the Hashamomuck Beach and Truman Beach barriers and cutting bluffs into the Browns Hills reach of the moraine. Across Goldsmith Inlet, northeasterly currents deposited the main spit, but modern drift is piling sand against the west side jetty.

Cliff erosion is extensive westward to Mattituck Inlet, where a sand bar once forced Mattituck Creek to the west. But the barrier development is now offset by sand accreted to the west side jetty by easterly drift. Bluffs cut in the moraine dominate the coastal landscape west of Mattituck Hills. Near Roanoke Landing sand drifted around submerged wrecks appears to be moving to the east, and at Fresh Pond Landing, a marsh occupies the drainage basin sealed off by the west to east barrier. At Wading River, an east to west sand bar has forced the tidal channel to meander westward across the salt marsh. North of Rocky Point and Sound Beach, the Roanoke Point Moraine is well-cliffed but with little evidence of longshore currents. The east to west Cedar Beach at Miller Place forms the barrier across Mt Sinai Harbor. Deposition of sediments in the harbor has resulted in tidal flats and marshes. The complex of spits defining Port Jefferson Harbor show both east to west and west to east currents and sand accumulation. At Mt Misery Point flow is to the northwest with a large washover widening the sand bar on the south side of the point. On the west side, Old Field Beach has grown to the east, and a sand bar isolates Flax Pond. From Old Field, the Roankoke Point Moraine is projected westward to its probable convergence beneath Smithtown Bay with the Sands Point Recessional Moraine from the west. While the interlobate position may be easy to infer, the submerged morainal deposits are no longer obvious. At this point, the geologic story of eastern Long Island meets that of western Long Island; the latter is related in a companion volume, Western Long Island Geology.

## SEA LEVEL RISE

As indicated in the preceding section, eastern Long Island is at the mercy of coastal currents that heavily erode the headlands and barrier beaches. Sand is being trapped at the many groins and jetties positioned along the coast. But the net loss of land has become a reality. We are increasingly aware of the erosion of the coast as bluffs and dunes recede during each successive storm. Seasonal erosion of the beaches exceeds the natural resupply of sand, and seaside communities are increasingly endangered. We might wonder whether some new natural component has arisen to disrupt the balance of beach processes. Actually, a couple of powerful forces have asserted dominance over the dynamics of the coastal zone.

For one, people have made a major impact on the coast, building in sensitive areas, dredging the sea floor, and rearranging the landscape, with little idea of geological and environmental consequences. Next, the world-wide effects of changing climate, supplemented by destructive environmental practices have impacted modern coasts. There is considerable scientific evidence to show that global warming , underway since the end of the "Little Ice Age", a natural cold climate interval that lasted from the fifteenth to the nineteenth centuries, has been accelerated by an increase in carbon dioxide in the atmosphere, a side effect of the extravagant use of fossil fuels since the industrial revolution. Higher levels of gases and particulate matter in the atmosphere have heightened the "greenhouse effect," the mechanism by which the heat reradiated by the earth is trapped in the atmosphere. As a consequence, warmer climate has caused glaciers to melt and the level of the sea to rise. The slow but steady rise of the sea is the underlying cause of our vanishing shoreline.

One prominent theory of the cause of ice ages and the alternating warm intervals deals with cyclic and differential heating and cooling of the earth due to the changes in the tilt of the earth's axis, the shape of the earth's orbit around the sun, and the distance of the earth from the sun. While the orbit varies from nearly circular to an elongate ellipse, the optimum-for-warmth position has the earth's axis vertical and the earth closest to the sun. It takes about ninety-six thousand years for the earth's orbit to change from elliptical to circular and back, about the same duration as a major glacial and interglacial cycle. The axial tilt cycle takes about forty-one thousand years, about the same as a glacial subcycle, with half of that time, twenty thousand years, for the interval from the last glacial to the present warm episode.

Along with these comparable astronomical events and terrestrial climates, short-term, cold-warm cycles are attributed to the exchange of oceanic heat from the southern to the northern oceans via deep-water ocean circulation. The onset of Pleistocene glacial cycles, overall, is thought to be related to the uplift of the western North American plateaus during the Tertiary Period. The rising highlands had the effect of altering global atmospheric circulation. To all of these interacting agencies we can add atmospheric pollution and the increasing load of carbon dioxide and other gases, plus the dust from occasional volcanic eruptions. The result is an enriched greenhouse effect that has increasingly warmed the atmosphere and intensified the interstadial climate. Researchers have already noted increases in the global heat budget, changes in plant and animal distributions, particularly in the life of the southern ocean, and most importantly the increased melting of glacial ice that has contributed to the rise of the sea.

The major effect of higher sea level on a worldwide scale will be the inundation of the coasts. Coastal cities will be flooded; beaches, dunes and some islands will disappear, and river mouths will be more deeply drowned in sea water. Some tide gauges show mean sea level rising about one eighth of an inch per year, or about one foot every hundred years, a rate that matches the measured rate of salt marsh growth in this region. In three hundred years sea level may have risen normally as much as three feet. In the future we may have to add an additional increment due to global warming.

What does this much sea level rise look like to us? Try a simple test at the beach when the tide is high: imagine a water level about four feet higher, then project that level across the dunes, bluffs, and seaside buildings. It is not hard to see how much of the shoreline would be under water. On a topographic map, the new shoreline can be approximated by drawing an additional contour line between the sea level shown and the first contour line representing an elevation of ten feet above the mean sea level datum (Figure 19). What happens to the coastline on the map? The new shoreline would dissect eastern Long Island: beaches and marshes would be eroded and submerged, dunes erased, islands created, and the sea would be up against and cutting into the bluffs. The jetties, built to protect the inlets and harbors, would be awash. Of course, this rise of the sea is slow going like most geological processes. We might not notice the changes until their effects become critical, but over time, the slow rise of

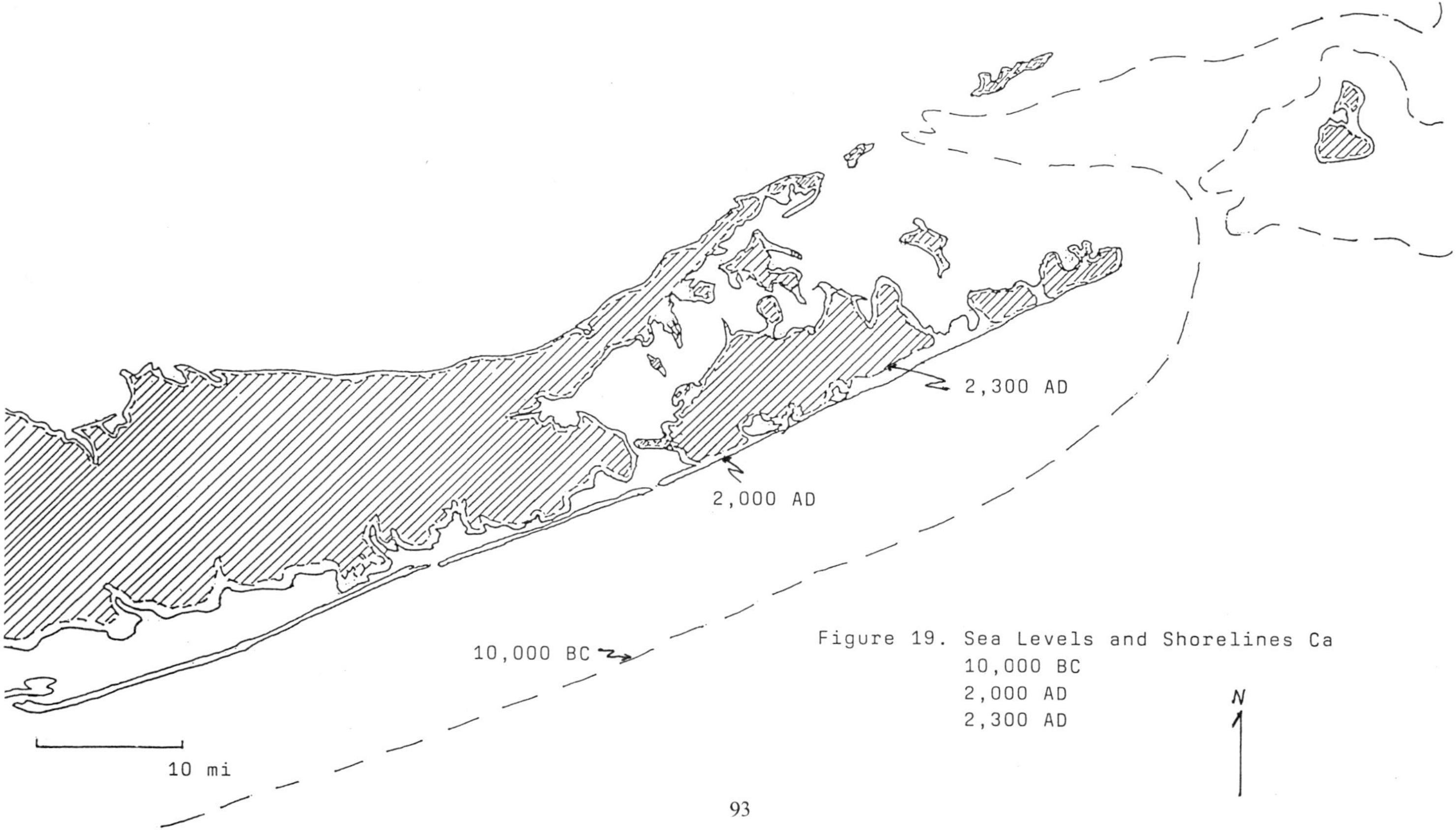


Figure 19. Sea Levels and Shorelines Ca
10,000 BC
2,000 AD
2,300 AD

the sea combined with erosion will be felt further and further inland. This model assumes the fragile coastal features will vanish-- in time somebody else will have waterfront property.

## EROSION AND MAINTENANCE OF THE SOUTH SHORE

The south shore barrier beaches are dependent for their existence on the movement and supply of sand from east to west. Recently, most beaches have been conspicuously eroding due to a number of interacting factors: frequency of major storms, increased slumping of the headlands and erosion of sediment source areas- due in part to sea level rise, loss of sand offshore and starving of downstream beaches as a consequence of construction of pairs of jetties to stabilize inlets and groin fields to protect small segments of the barriers. Consequently, the barrier beaches are responding through erosion and thinning, at a general rate of more than one foot per year east of the jetties, and as much as ten feet per year west of the jetties. There is an abnormal build up of sediment adjacent to the jetties and groins and migration of the barriers landward due to sand washovers. Two jetties at an inlet cause increased sedimentation by focusing tidal deposition between them and creating tidal sand bars in the channel.

From the Hamptons to Fire Island numerous natural, but temporary, inlets have existed in historical time. There are now only a few inlets, and these are protected by jetties. The east, or up-drift, jetty traps sand that would eventually close the breach. Shinnecock Inlet was cut by the hurricane of 1938 and has been protected by jetties since 1952. Moriches Inlet, opened in 1931, was also not stabilized until 1952. At Fire Island Inlet, a jetty nearly one mile long was completed in 1941 at Democrat Point. Prior to the jetty, the barrier beach had migrated about five miles to the west. The Fire Island Lighthouse, built in the early 1800's to mark the inlet, is now almost six miles east of the inlet where a small remote signal guards the channel.

The long-term effects of this artificial beach maintenance regime are apparent because many beaches have thinned to the point that houses, roads and sanitation systems are threatened. They are dramatically revealed after a storm. In one hurricane-like storm in the winter of 1992, storm surge ran eight feet above normal. Over one hundred homes were destroyed between Westhampton Beach and Fire Island, and the barriers were breached in several places. The unprotected western end of Westhampton Beach was virtually flattened. Dune road was under water and two new inlets had been cut through the barrier. Millions of cubic yards of sand had to be imported to fill the new inlets and restore the beach.

The Westhampton Beach situation developed because a field of fifteen groins had been constructed to trap sand and protect the eastern end of the barrier (Figure 20). The groins were effective and the protected part of the barrier looks just as permanent as the rest of suburbia. The western end of the beach, with no groin field, was open to erosion by the sediment-depleted longshore current. Over the years the barrier has diminished, annually losing ground and suffering inundation by the sea. A series of storms in the early 1990's finally made the area west of the last groin uninhabitable. With the persistence of sea level rise and coastal processes, future storms will undoubtedly cause similar damage. Some environmentalists believe that beach restoration is a wasted effort and that beaches should be left in a natural state. However, current beach maintenance practices demand rebuilding the beach to its former levels. Dredging, importing sand, construction of additional groins, and periodic replenishing of the eroded sand have become the preferred techniques. Certainly, future storms will wash away the new sand; but between storms landowners will be able to return to the beach and rebuild their homes.

a) West of the Last Jetty (in foreground)

b) West of the Jetty- end of the protected area is in left foreground

Figure 20. Scenes From West Hampton Dunes After Storms

## HYDROGEOLOGY AND WATER RESOURCES

Several years ago at a meeting of concerned eastern Long Island residents, a visiting geologist cautioned against overdevelopment and the risk of contaminating the Island's waters, including its restricted supply of fresh water. The scientist reviewed the hydrogeology of the region and the nature of the subsurface water resource, emphasizing that the primary source of fresh water was rainfall and concluding with a discussion relating water supply, sanitation and population. After a respectful time, one of the more prominent political figures in the audience rose, thanked the geologist for his presentation, and then responded to the scientific evidence with what had been the gospel of east end water lore for centuries: everyone knew there was an underground river from northern New England. This river flowed deep beneath eastern Long Island and brought an unlimited supply of fresh drinking water from the mountains of the north to the local residents.

Unfortunately, this bountiful water supply is only a long-standing myth. All of the fresh water for Long Island originates as rainfall, the sole source of the water, which has been soaking into the soils and strata since the glaciers melted away twenty thousand years ago. There are few rivers; none of them originates outside of the Island or discharges enough water to supply the growing population. The surface water is stored ponds, streams and wetlands, and below the surface it occupies the pore spaces of sedimentary layers. Where the shallow sediments are saturated with water, a groundwater surface or water table forms, while deep below the surface, water-saturated layers of sediment become natural reservoirs of water known as aquifers.

Surface water recharges the underground formations and also carries a variety of pollutants into the soil. In fact, the aquifers have been subjected to many abuses due to long-term "out of sight-out of mind" waste-disposal practices. In recent years legislation has been enacted to protect aquifer recharge areas from the dumping of commercial and industrial pollutants.

The public water supply generally comes from the deeper aquifers in layers of Cretaceous sands because in many areas the water in the glacial deposits near the surface has already been contaminated. Where it has not been spoiled, the glacial aquifer is easier to access and may still provide water for private homes. But careful monitoring of this resource is required. The water in the deeper Cretaceous beds directly over bedrock is least susceptible to surface pollutants, but deep wells may encounter water rich in naturally-occuring iron oxide, clay or salt water. This raw water often requires some special chemical treatment to remove contaminants and bring it to drinking water specifications, but at this time removing salt from the aquifer water supplies is not economically feasible.

## RAINFALL

In any case, water purification can not increase rainfall, and when use exceeds recharge, as it does in so many growing communities, the public water supply can become inadequate. Both surface and subsurface water sources depend on rainfall, and in drought years many wetlands and many private wells may dry up. Water wells also rely on the water transmitting characterisitcs of glacial and preglacial strata. The sediments in these layers are mainly unconsolidated sediments that vary in grain size, sorting, compaction and chemistry. It will be worthwhile to review some surface and subsurface water regimes.

Rainfall has averaged around forty-five inches per year over the long term, but it can drop as low as thirty inches per year during droughts. On a monthly basis, the average rainfall ranges from a low of two and one half inches to a high of four inches. The low, or water deficient months, are from June to September, coinciding with the seasonally high population, higher water use, and increased rates of evaporation rates of surface waters.

Rain that falls on the surface tends to seek lower elevations. Powered by gravity, water running over the surface in thin sheets begins to cut into softer soils, gradually forming gullies. Gullies merge to form stream channels, and these slowly widen and deepen. Ultimately the larger streams coalesce to form rivers, many of which flow to the sea. On eastern Long Island, some streams may naturally feed ponds and marshes, while others have been dammed to form reservoirs and mill ponds. Surface waters also pick up and carry with them quantities of sediment loosened by weathering and mass wasting. Much of this sediment is eventually redeposited in the channel and some reaches the coastline. After a heavy rain, remobilized sand flows into roadways, and finer sediment forms plumes of muddy water in ponds. About half of the rain seeps into the soil and supplies the groundwater. A small portion of this is taken up by plants. Water returns from the surface- ponds, streams and wet ground- to the atmosphere through the process of evaporation. Water is also given off by plants by evapotranspiration, that is, plants take groundwater in through their roots, use it internally, and then they expel it through specialized pores into the air.

STREAMS AND RIVERS

About half of the rain water reaching the surface runs off in streams. On eastern Long Island there are numerous ponds and small lakes, such as Lake Panamoka near Wading River and the centrally located Lake Ronkonkoma, a kettle lake that is embedded in the Ronkonkoma Moraine. In fact, many of ponds and lakes are glacial in origin, but some of the ponds lie along stream channels and are man made. In the South Fork, chains of ponds (pater noster lakes) occupy kettles in meltwater channels south of Sag Harbor. There are also many lakes that were formed by barrier beaches cutting across stream channels along the coastlines. A number of shorter stream channels drain the outwash plain and empty into the bays. The stream channels and bays have been drowned by rising sea level and make up a network of coastal estuaries that resemble a large and complex saline lake.

Eastern Long Island has many south-flowing streams, like those mentioned above, that appear to have formed on the southern edge of the outwash plain as postglacial drainage basins. However, only a few permanent rivers drain the upland and flow to the sea. In studying the outwash plains, both south of the terminal moraine and between the terminal moraine and recessional moraines, there are a few streams large enough to be called rivers. There are also valleys occupied by stream channels that are dry or only seasonal. A few broad valleys have south-flowing streams that are too small to have eroded the valley. Geologists refer to these as underfit streams. The valleys in which they flow are actually ice age relicts, eroded by much larger streams that carried meltwater southward from the melting glacier and incised broad channels into the outwash plain. These valleys slope southward toward a much lower, glacial-age sea level that existed far out on the continental shelf at the time of valley formation. Now the undersized streams meander on narrow flood plains through steep-walled and meander-scarred, broad valleys.

The few permanent rivers in this region include the Carmans River and the Peconic River, the largest of the fresh water sources (Figure 21). The Carmans River begins in a pond near Middle Island and follows a sinuous and often dammed-up channel for over nine miles toward Shirley, and then an additional four miles southward as a tidal channel. The west to east flowing Peconic drainage lies between the moraines, the Ronkonkoma Moraine to the south and the Roanoke Point Moraine to the north. The river flows over glacial terrain that includes outwash sands, proglacial lake clays and north to south chains of ponds. The Peconic rises in a pond in central Long Island south of the village of Ridge and begins a relatively long, fifteen-mile, meandering journey eastward to Riverhead. Along the way the river is fed by chains of ponds south of Lake Panamoka and south of Calverton. It is dammed for reservoirs like Lake Peconic, and some of its wetlands have been converted to cranberry bogs, supporting a local industry. On entering Riverhead the river widens and becomes tidal, flowing into Peconic Bay.

The largest north-flowing river to affect this region is the Nissequogue, which shares a low divide near the Ronkonkoma Moraine with the south-flowing Connetquot River. The lowland they both seem to share appears to have been a glacial meltwater channel that once carried water from the ice near the Roanoke Point ice margin and later from proglacial lakes in the vicinity of Long Island Sound southward through the terminal moraine. These major rivers

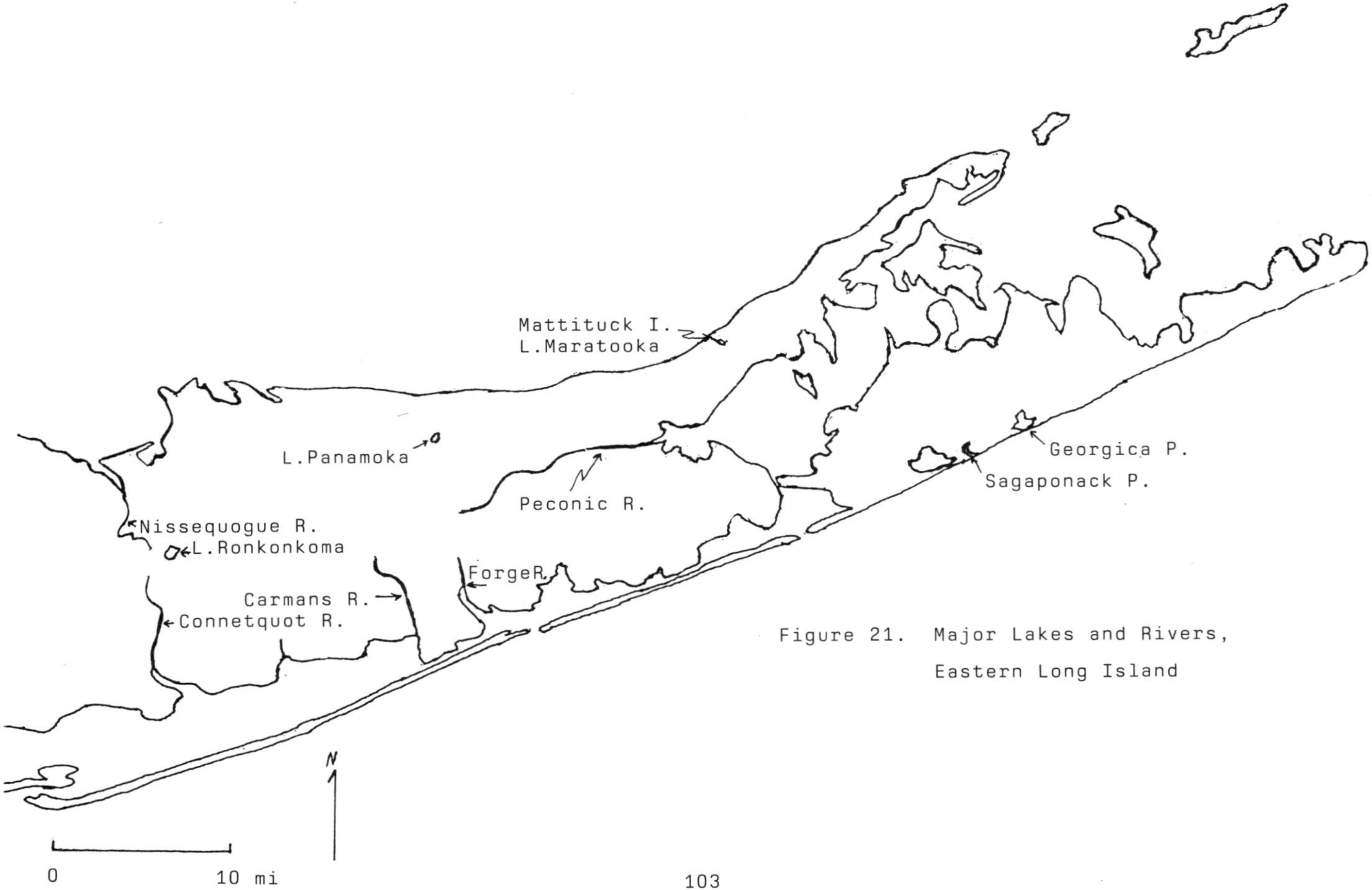


Figure 21. Major Lakes and Rivers, Eastern Long Island

of eastern Long Island drain large land areas; the Peconic and Carmans have combined drainage basins of over one hundred fifty square miles and the potential of carrying a volume of water approaching five million cubic feet per day.

## THE GLACIAL AQUIFER

Water that infiltrates the soil moves into the open pore spaces between the sediment grains in the glacial deposits. These deposits vary in their ability to hold or transmit water. As explained earlier, the glacial drifts are the result of two separate, late Pleistocene glaciations. They both include tills, outwash and lake beds and incorporate structural deformation due to ice shove. The superimposed drifts appear as two tills separated by outwash. The older, lower till contains clay and is compact, more like a basal or lodgement till. It is less able to transmit water or allow water infiltration and is categorized as being relatively impermeable. Near the surface it supports ponds, wetlands and springs, and at depth it acts as a confining layer for groundwater. Soil water accumulates in the sediments above the till- mainly the outwash of the younger drift- to create a groundwater table and a saturated and permeable aquifer. The younger, upper drift, which forms the morainal topography is sandy and less compact. It readily transmits water and provides a source of near-surface water above the lower till. It also serves as the porous and permeable unit of the glacial aquifer. The upper till ranges from basal till in some areas to thin sheets of meltout till, flow till or layers of till stones elsewhere.

The glacial aquifer in eastern Long Island is a hydrologically continuous, general zone of saturation located in the vicinity of sea level (Figure 22). Its surface is the groundwater table.

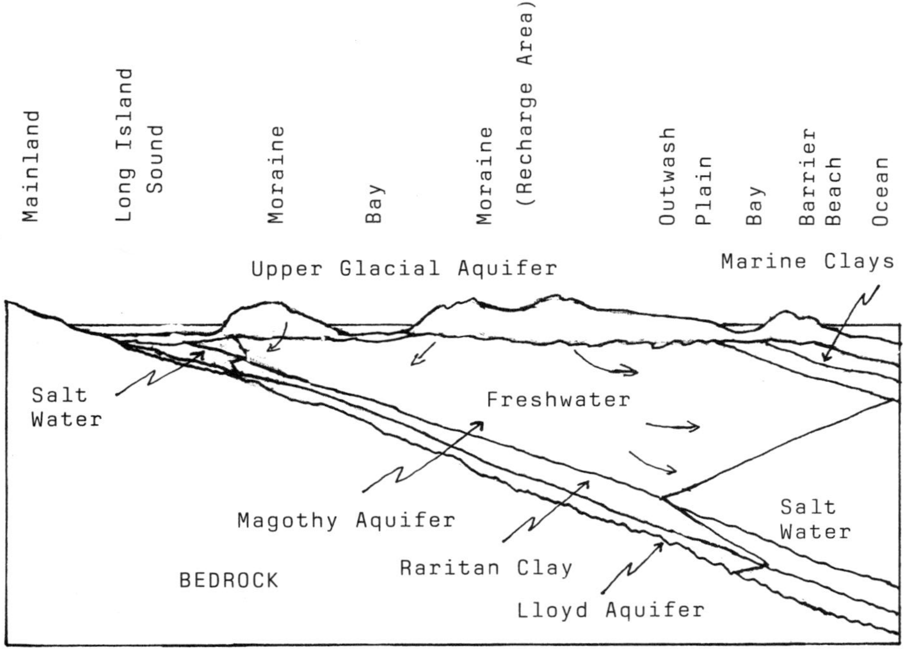


Figure 22. Eastern Long Island Aquifers (Flow Lines ⟶ )

Near the coastlines, freshwater in the glacial aquifer may be displaced by salt water. Groundwater is also stored at higher elevations above the regional water table, perched unconfined on top of impermeable layers of the lower till. Water seeping out over the till surface forms springs and ponds. Seeps may provide sustenance for aquatic life, and springs with their mystical healing properties-but questionable purity- have long been popular sources of drinking water. Under small islands and barrier beaches, fresh water may exist in the coastal sediments. As a general rule, a fresh water lens can extend below sea level to depths up to forty times the height of its water table above sea level (Figure 23). The barrier beach aquifer is separated from the outwash plain portion of the upper glacial aquifer by sediments saturated by salt water. This coastal fresh water lens is also precariously situated above salt water and below the septic disposal systems of coastal homes and businesses.

## The Deep Aquifers

The glacial deposits are underlain by layers of Cretaceous sediments of both marine and coastal plain origin. Most of these sedimentary layers are porous and permeable to some degree. Layers of low permeablility, which usually have high clay content, impede the flow of water, and form the confining layers, or aquicludes, for the aquifers. Aquifers include those layers that are composed mainly of granular sediment and effectively store and transmit water (Figure 22). Beneath Long Island, the aquifers are named for their geologic counterparts in the New Jersey coastal plain. Although additional strata have been identified the aquifers carry the collective hydrologic name. Thus, the Magothy aquifer makes up the bulk of the wedge

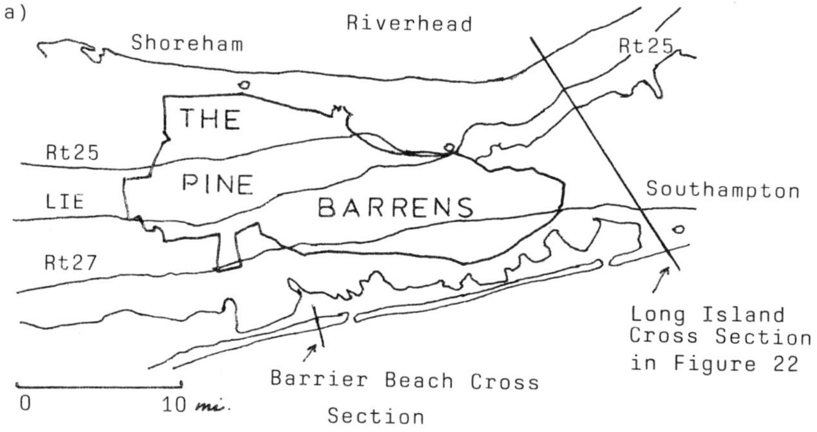


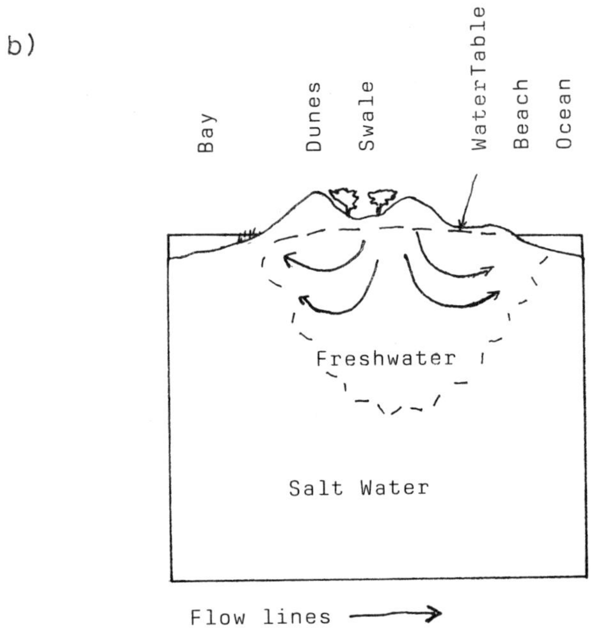


Figure 23. a) Eastern Long Island: Aquifer Cross Section Locations and the Pine Barrens

b) Hydrology of the Barrier Island Aquifer

of sandy and clayey sediments below the upper glacial aquifer and above a clay unit. It correlates with the Magothy Formation and younger Cretaceous strata in New Jersey, while the underlying clay, silt and sand sequence is matched with the Raritan Formation.

In southeastern Long Island, the Magothy aquifer is partly confined above by the Gardiners Clay, and generally confined below by the Raritan clays. Groundwater flows downward into the Magothy from the upper glacial aquifer under much of Long Island. Below the Raritan lies the Lloyd aquifer, a layer of coarse sand directly above bedrock. The three-part nature of this geological sandwich enables well drillers to create a water well by drilling a hole from the surface into a permeable, water-bearing stratum. Between the surface and the lower aquiclude, a casing pipe is installed with a porous screen in the pipe at the level of the aquifer. The well is then sealed above and below the screen. Now only water from the aquifer can flow into the pipe where a pump can deliver water to the surface. The casing of the well is also sealed against the sediments up to the surface, extended above the ground and capped to prevent contaminated surface water from entering the well.

Water is transmitted through the aquifers depending on the nature of the materials and the slope of the layers. Rates of flow can reach as high as a few hundred feet per day and involve millions of gallons of water. This means that groundwater does not take up permanent residence in the subsurface. Porous strata slowly leak water from bluffs, surface seeps and springs, and below sea level into the ocean. It also recharges the deeper aquifers. Throughout the region, water recharge from a soaking rain can add tens of millions of gallons to the aquifers.

Groundwater flow in the regional zone of saturation generally follows the configuration of the topography, so that recharge to the aquifers begins at the groundwater divides, follows slopes and enters the deeper aquifers through zones of permeability (Figure 22). Flow can be complicated by the shape of the till surface and by the chaotic arrangement of irregular blocks of till and masses of Cretaceous strata that have been thrusted by the glaciers. Where superimposed, these thrusted units can result in poor water recovery from wells.

The Cretaceous sediments may also be rich in iron oxide which can leach into the groundwater. Water flow in zones of low permeability may follow pipe-like channels around which oxides can be redeposited to form tubes encrusted with iron minerals . In outwash, the iron oxide cement can create a conglomerate of sand and gravel that has been mistaken for an ancient glacial "red till." The red conglomerate and iron-oxide piping structures have often been exposed by coastal erosion, and they stand out from the bluffs due to their greater resistance to the waves.

## GROUNDWATER RECHARGE AREAS

While recharge to the groundwater table is general, upland areas are most effective in supplying water to the aquifers. Consequently, it has become essential to recognize and designate actual groundwater recharge areas in order to establish programs for protection of the freshwater resources. Because groundwater recharge to the deeper aquifers begins in the uplands, the protection areas in eastern Long Island include the moraines, from the Interlobate Moraine Zone near Melville eastward to the terminal moraine and recessional moraines of the east end.

The Ronkonkoma Moraine in central Long Island forms a divide with surface water flow and infiltration into the subsurface southward to the outwash plain and the Pine Barrens and the underlying aquifers, and northward into the Peconic River watershed, the upper glacial aquifer and the deeper aquifers. The terminal moraine continues into the South Fork as the Shinnecock Hills Moraine and the Amagansett Moraine. The thick layer of porous and permeable upper drift forms the principal aquifer and supplies some recharge to the Cretaceous strata below.

The Roanoke Point Moraine along the north shore recharges water southward into the Peconic drainage as well as into the aquifers, but with only a thin layer of glacial sediment available for groundwater infiltration. The moraine and the outwash plain here are not as thick or as wide as the Ronkonkoma Moraine and the southern Long Island outwash plain and are surrounded by salt water bodies (the hydrologic cross section in Figure 22 applies to this situation as well). Thus, the glacial aquifer is a more limited resource and the North Fork water supply is less abundant than in the central part of the Island. As noted, the freshwater beneath the North and South Forks as well as the barrier beaches is not only dependent on local rainfall but is also subject to salt water intrusion.

## THE PINE BARRENS

Perhaps the best groundwater resource in eastern Long Island is in the sedimentary layers beneath the Pine Barrens, a region of pitch pine and scrub oak forest established on sandy, glacial soil. While the Pine Barrens encompasses over one hundred thirty thousand acres of land between Coram and Riverhead from west to east, and from Shoreham to Westhampton north to south, only about half of the land is underdeveloped. Recharge from the developed portions seriously compromises the quality of the water in this part of the glacial aquifer. Real estate developers see the Pine Barrens as attractive and available open land for new housing and shopping malls. But many people are impressed with the potential of the open land for its ecological significance and its value as a recharge area and a major regional water resource. They believe it should be protected from development and pollution by being designated a special groundwater protection area (Figure 23).

The Pine Barrens Protection Act attempts to accomplish this by creating a fifty-three thousand acre, limited-development core preservation area and a Pine Barrens Commission to evaluate development projects. Previously, project proposals were considered in a piecemeal fashion. Individual projects for development in the Pine Barrens were considered by separate jurisdictions in terms of their local impact. Invariably, projects were approved because their local impact seemed insignificant. As a result, the Pine Barrens have slowly been whittled away. With the new Commission, proposed projects will be evaluated in terms of their potential impact on the entire region. This way the combined impacts of multiple projects on the region will better understood.

## GROUNDWATER QUALITY

The groundwater in eastern Long Island can be somewhat acid and corrosive. Analyses of water samples show distinct trends in acidity and chloride content due to salt water encroachment, as well as increases in iron, nitrate, sulfate, and several other compounds. When demand for fresh water increases, higher pumping rates pull the fresh water level down, and allow salt water, with chloride levels between one thousand and ten thousand milligrams per liter (mg/l), across the zone of mixing or diffusion. Because of salt water intrusion, water wells close to the coast tend to produce water with high salt content as salt water replaces fresh water in the sediments (Figure 24). Chloride levels of around thirty mg/l are considered normal, but close to the shoreline chlorides above one hundred mg/l are commonly measured. High pumping rates in the Magothy Aquifer have allowed salt water to intrude far inland from its pre-Colonial position.

Levels of iron over four mg/l may be associated with water derived from iron-rich Cretaceous sediments either from the subsurface Cretaceous aquifers or from perched water in the ice-shove masses of Cretaceous material. High iron and sulfate may occur in ponds where biological activity causes low oxygen levels. This can result in reducing conditions and precipitation of iron compounds. Under these conditions, bacteria fix iron and sulfur to clay minerals to form the precipitates.

Sources of groundwater pollution are tied to discharge of residential and commercial waste water, leaking underground oil and chemical storage tanks, defective septic systems, landfill leachate, and use of agricultural and landscaping chemicals. Common toxic chemical contributions are dry cleaning fluid, aviation fuel, automotive fluids, such as oil, gas, lubricants, battery acid and antifreeze from junked cars, and nitrates, herbicides, and pesticides from farms, lawns and golf courses. Illegal dumping of waste water, household and industrial chemicals, fuels, lubricants, detergents, and garbage into septic systems, sewers, water bodies

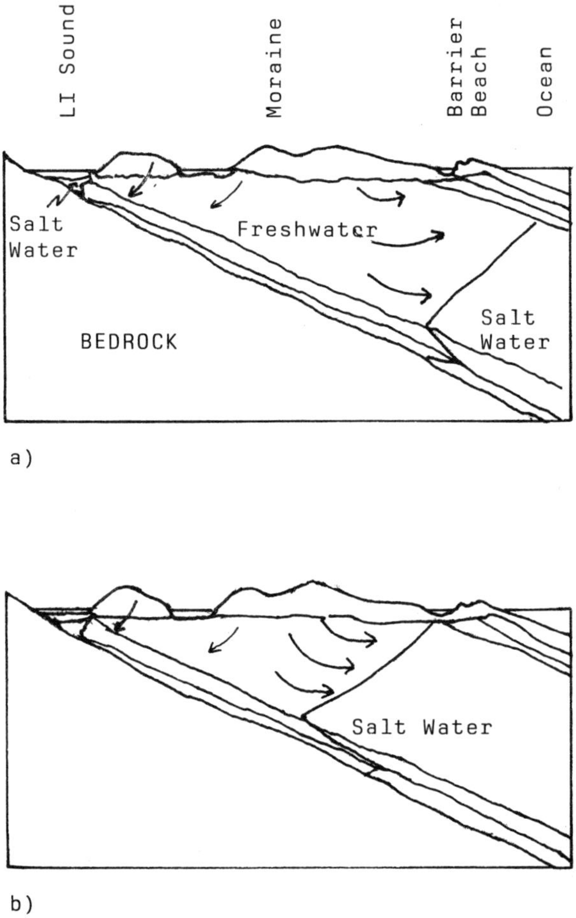


Figure 24. Saltwater Intrusion a) before extensive water use b) after (Flow Lines ⟶ )

or just "behind the barn" is a hard to regulate activity and a leading source of groundwater contamination. Leaching of these pollutants into the aquifers can be a long-term process with serious impact on water quality. Many of these compounds cannot be filtered out and are not biodegradable, so that groundwater will remain polluted indefinitely.

In regions of sole source aquifers, like rainfall dependent eastern Long Island, slow groundwater recharge and water transmission means that toxic chemicals are only slowly flushed out of the system. They may remain in the aquifer and may even become concentrated over time. Many former landfills have generated toxic groundwater plumes that will affect water quality for years. Higher than normal amounts of colliform bacteria, iron, nitrate, sodium, sulfate, lead, zinc, manganese, chromium, and automotive fluids, as well as high acidity and electrical conductivity-often as much as thirty times normal- have been associated with groundwater near landfills. Due to geologic constraints, plumes can migrate into aquifers following permeable zones in the sediment layers or flowing over less permeable zones. To prevent rainfall from supplying the leachate water, landfills must now be capped with layers of plastic and clay.

## WATER CONSERVATION

In view of the pollution potential, eastern Long Island's water supply, although proclaimed adequate, equivalent to nearly a billion gallons per day according to United States Geological Survey-Water Resources Division studies, is continually monitored. Certainly, protecting this water supply is one of the best ways to conserve it. Point sources of contamination are regularly identified and some have been cleaned up. Antipollution laws have been enacted to eliminate the problems of leaking storage tanks and defective septic systems and to prosecute illegal dumping. Improved double-walled tanks for both below and above ground use have replaced buried single-walled tanks in gas stations, commercial properties and residences. In-ground, steel septic tanks are yielding to large capacity concrete vaults, and some individual septic systems are being replaced by community sewage lines and sewage treatment plants. This strategy lessens the impact of waste water entering the water table locally, but defeats the conservation aspect, because most sewage treatment plants discharge treated water into the ocean.

The high per capita water use, estimated variously between one hundred and one hundred fifty gallons per day, calls for some kind of controls on individual water use. These can range from the solely individual exercises ("Please turn off the water"), to regulations ("Lawns may be watered on alternate days"), repair and replacement of leaky pipes and valves, water caps and water meters with escalating charge for high water use, and even a greater attempt to clean up the upper glacial aquifer, a very large but highly abused and under-regulated water resource. Water caps have been employed and have met mixed reactions in acceptance. The fear of higher governmental regulation on the amount of water pumped from an aquifer by local communities is as great as that associated with any federal jurisdictional that appears to usurp states rights, even though the aquifers are shared by all communities. The water supply future may involve use of tertiary treated waste water and desalinization of sea water to supplement a dwindling and polluted resource. But elimination of sources of contamination and reclamation of polluted aquifers could be effective with the cooperation of all parties.

Furthermore, because the regional water supply flows slowly seaward and is essentially lost in substantial quantities each year, additional water storage upstream could be a viable alternative. In any case, any realistic conservation strategy should be employed to create and protect an unpolluted, steady state water supply.

## INTRODUCTION TO FIELD TRIPS

In the context of geologic time- over four and one half billion years- the formation of eastern Long Island is barely a footnote. It took only about two thousand years, between twenty-two thousand and twenty thousand years ago for glacial advance and deposition of the terminal moraine and deglaciation and deposition of the recessional moraines. From twelve thousand years ago to the present, sea level rise and coastal erosion have shaped the moraines and created the characteristic, fish-shaped Island geography. Consequently, the variety of glacial and postglacial features, ranging from ice-contact deposits to relict landforms are readily available for study. These features are well represented and readily accessible in coastal bluffs and on the surface, and they are easily seen from roadways, beaches and trails.

A few typical excursions to various locations in eastern Long Island are described in this section. Each trip includes a variety of interesting geologic features and scenic views. The descriptions include maps, directions, explanations and sketches (Figure 25). Distances along the routes are approximated in tenths of miles, and geologic sites are keyed to place names referenced on the United States Geological Survey 7.5' topographic quadrangle maps. Individual trips are shown on selected map segments. These are catalogued in the index of topographic maps (Figure 26).

### CAUTIONARY NOTES

Because eastern Long Island is already greatly diminished by the unrelenting forces of nature, please do not collect samples of rocks, plants or insect species. And because this

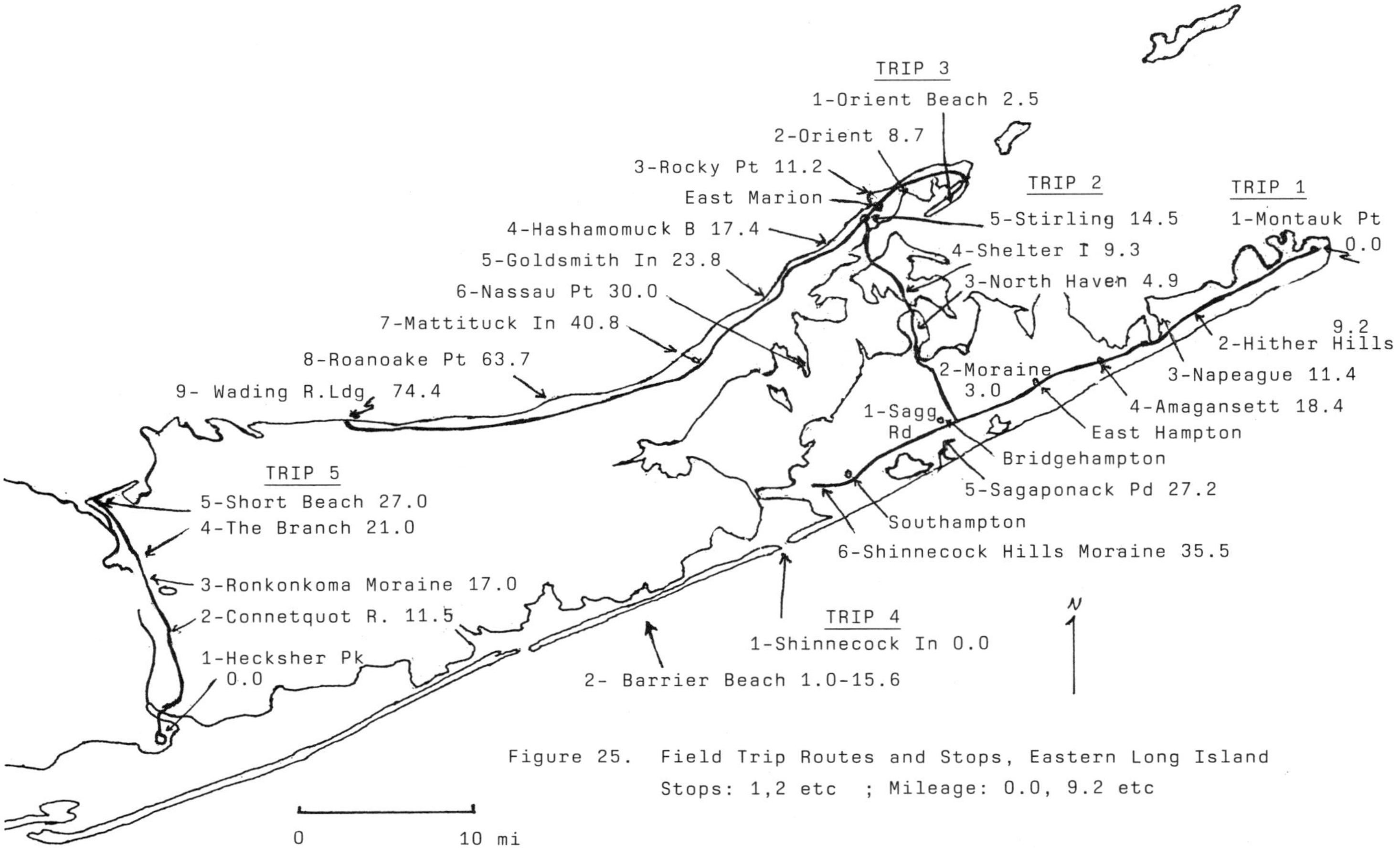


Figure 25. Field Trip Routes and Stops, Eastern Long Island
Stops: 1,2 etc ; Mileage: 0.0, 9.2 etc

Figure 26. Index of Eastern Long Island Topographic Maps

region is the breeding ground and is on the migration route for many species of birds, please do not disturb nesting sites or enter posted wildlife areas. Be additionally aware of the presence of deer ticks and other disease vectors and take the appropriate precautions.

**Field Trip Geology**

The geology of eastern Long Island, as seen in its coastal bluffs, gravel pits, and landfill excavations is characterized by glacial deposits and deformation and erosion of the two drift sheets. Various exposures of these features allow us to examine the results of glacial and postglacial processes, often in open cross sections in the direction of ice movement-roughly north to south- and across the grain of ice advance from west to east.

The major stratigraphic units are the upper (younger) and lower (older) drifts, both of which contain tills, outwash and lake beds. They are distinguished first by their relative position, normally the younger, upper drift is superimposed over the older,lower drift. Color is the next important consideration: the older drift is darker and in shades of gray; the younger drift is lighter and in yellow-browns. The colors of both drifts depend on mineral content and provenance.

The upper drift is rich in granitic rock debris owing to its northerly source areas in granitic terrane. The till stones and erratics, combined into one category, are designated here as "granitic." This is an arbitrary grouping that includes granite, granite gneiss, pegmatite and quartzite, rocks with similar color and mineral content. The lower drift, which is more readily observed in the South Fork, has more of the dark-colored rocks from the mainland to the northeast, including dark-colored gneiss, schist, phyllite, metaconglomerate and coal. In this context, the field trip descriptions may cite "granitic gravel" or "dark-colored till stones" as suggestions of the presence of the upper or lower drifts.

The oldest sedimentary deposits in the region are beds of sand, clay and gravel of Cretaceous age, which may appear as masses of sediment thrusted into the moraines by the last glacier. These Cretaceous clasts were moved by the glacier from the edge of the uptilted strata now submerged beneath Long Island Sound north of eastern Long Island. Next, the lower drift unit is comprised of till, often a variety known as basal or lodgement till deposited and compacted beneath the glacier. The lower drift may be in place where it was deposited, or deformed, ice shoved and detached from the main body of the drift. Masses of this sediment have been ripped-up, thrusted and engulfed in the outwash of the younger drift. Lake sediments in the lower drift include banded clays that appear as tightly-folded and thin-bedded sequences.

In the upper drift, outwash beds may slope away from the ice or are folded, steeply inclined, and intercalated with thrusted Cretaceous beds and lower drift. The most common, upper drift till is meltout till, which as the name implies has melted out of the ice. It may be partly washed by meltwater, but is still more or less unstratified. Meltout till may overlie outwash and varies in thickness from several feet to single layers of till stones. Tills interbedded with outwash may be flow till, a variety that flowed as a layer from the ice along with the sand and gravel, or beds of lake clays homogenized by thrusting.

The outwash consists of lenses of sand and gravel deposited adjacent to the glacier by meltwater streams. Fan-shaped deposits of outwash along the ice margin may coalesce into broad outwash plains like the outwash plain of Bridgehampton. Some sand and clay layers are deposited in proglacial lakes to form kame deltas or in lakes removed from the ice margin to form deltas. Due to glacial deformation, the complexity of geologic exposures is often hard to explain. Furthermore, deposits of varying thickness and structure gradually change in complexity because coastal erosion, slumping or new excavations alter dimensions and form, even though the relative position of the beds remains the same.

Postglacial, wind-blown silt, called loess, is found in thick layers covering the glacial deposits. The loess can be several feet thick and contain old soil profiles, or palesols, buried under recent vintage sand dunes. Beach deposits, excluding the flotsam, are the result of the interaction of coastal processes and the erosion of the bluffs. On beaches that lie below bluffs, we can find berms of coarse gravel, erratics, slump blocks with masses of sediment and vegetation, and slides of sediment from high up the cliffs. This eroding part of the coast is in the process of nourishing the beaches. The condition of beaches can also vary with the season or sequence of storms; sand won from the headlands may be pushed ashore by gentle summer waves, or massive erosion of beach sands, and dunes, and coastal structures may follow a storm.

The overland trips reveal many surface features resulting from glacial deposition and erosion. The sediments at the surface are likely to be those of the upper drift. The terrain of glacial deposition, erosion, and recessional features, such as meltwater channels, proglacial lakes and ice-contact deposits, is as distinctive as the hummocky topography of the moraines. The moraines form an upland with a crest or ridge line and slopes- the one facing the glacier is the proximal slope, the one facing away is the distal slope. In addition to these features, the landforms have also been modified by postglacial fluvial erosion and deposition and the formation and movement of sand dunes. These all combine to make the overland geology a fascinating complement to that encountered along the coasts.

## FIELD TRIPS: Routes, Stops, and Descriptions

### FIELD TRIP 1

Montauk Point and the Moraines of the South Fork (Figure 27). The starting point (Stop 1) for this trip is Montauk Point, the most prominent headland on eastern Long Island. If you are driving to Montauk Point, follow New York State Route 27 (Rt 27), the Montauk Highway. A recent Long Island road map is recommended. The topographic quadrangles will assist in locating specific features. Enroute there are a number of land forms that are easily recognized. Finding these features will help in understanding the geographic orientation of the moraines, and their position with respect to modern sea level.

Montauk Point is about twenty-two miles east of the junction of Routes 27 and 114 (Sag Harbor-East Hampton Turnpike) in East Hampton. Eastward toward Montauk, Rt 27 traverses the Amagansett Moraine, a segment of the terminal moraine, between Amagansett and Devon. Here the seaward extension of the moraine has been eroded away as a result of sea level rise, meltwater drainage, and coastal processes over the last twelve thousand years. Erosion has left an escarpment-possibly a wave-cut cliff- in the terminal moraine north of Beach Hampton. Subsequently, the accretion of sand bars formed the Beach Hampton plain. The beach from Beach Hampton to Napeague Beach, once eroded by meltwater has since been rebuilt by sand in the longshore current and has prograded to form a set of tombolos (the Napeague Isthmus) filling the late-glacial meltwater channel.

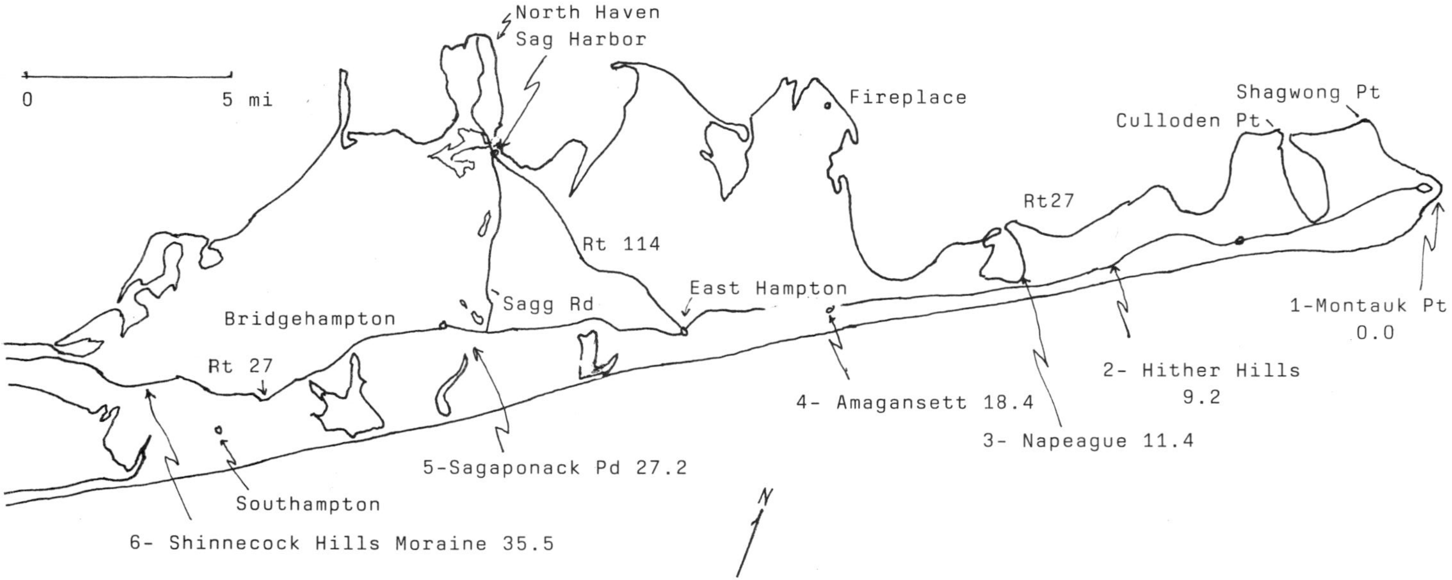


Figure 27. Field Trip 1 with Field Trip Stops and Mileage.

This meltwater channel once drained proglacial lakes, located to the north in what are now Gardiners Bay and Long Island and Block Island sounds. This river contributed to eroding the terminal moraine to the south and southeast.

East of Napeague in Hither Hills, the highway ascends the surface of a segment of the first recessional moraine to be deposited north of the eroded terminal moraine. The seaward extension of the terminal moraine now lies beneath the ocean, mainly as a lag deposit of glacial erratics. An exposure of the recessional moraine can be seen in the railroad embankment on the north side of the highway and east of Napeague Harbor. Here a forty-foot thick, kamic deposit is composed of outwash covered by sandy till. Montauk Highway traverses the moraine, mostly along the distal side, to Montauk Point where the road forms a loop passing the lighthouse and pavilion. Turn left at the gatehouse and park in the lot where convenient. Walk across the road to the pavilion terrace to check your location on the geographic map of Montauk and the adjacent region.

Stop 1. 0.0 miles. Montauk Point (Montauk Point Quadrangle, Figure 28). Follow the paved path on the east side of the pavilion toward the beach. Note the pond (a sag pond) in the depression on the south side of the path. This depression is a sag between the main slope and a slumped block of sediment adjacent to the beach. On the topographic map, several sag ponds appear along a fringe of slump blocks adjacent to the eastern beach. Each slump block has separated from the moraine and rotated downward toward the ocean along a water-lubricated, rotational fault, and each has the distinctive back-tilted and vegetation-covered, upper surface. Wave erosion has exposed the local glacial stratigraphy. The first slump just south of the path contains mostly upper drift outwash, overlain by a mass of deltaic, foreset sands. The delta

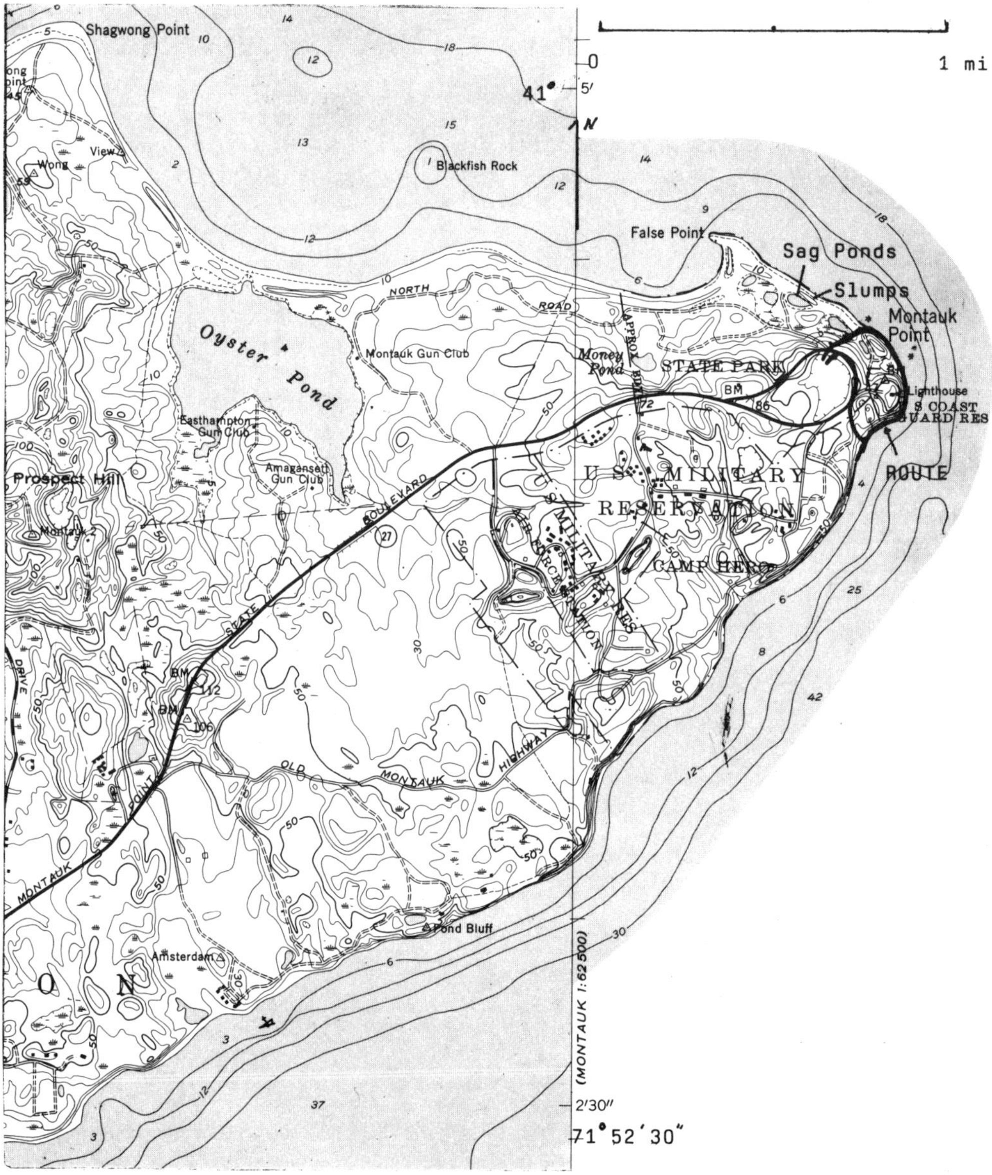


Figure 28. Montauk Point

was deposited in a proglacial lake that formed locally between the moraine and the glacier and is a typical feature of this environment. It is similar in morphology to other deltas found near ice fronts in former glacial lakes, and in the absence of radiometric ages or paleontological correlations, deltas do not represent time or marker horizons. The next slump block to the south reveals the fine grained, lake bottom sediments, probably from the same proglacial lake, in a sequence of rhythmically deposited, thin bedded silts and clays.

To the south, Turtle Hill, the hill on which the Montauk Lighthouse (in operation since 1797) stands, has the tilted profile of an incipient slump block (Figure 29). Because Montauk Point has been eroding at an alarming rate (as much as seven feet per year during recent stormy years), the United States Army Corps of Engineers has undertaken to retard erosion by armoring the Point with a retaining wall made of large blocks of crystalline rock. A terrace of these rocks enables visitors to walk around the point. From the terrace observe the additonal construction of steps on the vegetated upper slopes. The risers are made of planks that support the soil in the steps. A variety of plants are growing on the flat surfaces, and their roots also help to hold the sediments in place. Both the rock wall and the steps have slowed the erosion process, but the energy of storm-driven ocean waves is great enough to dislodge the rocks and surge upward onto the vegetated slope. Unseen beneath the wall, currents winnow sand away and can ultimately cause the wall to collapse. Periodic inspection and maintenance should prevent major damage.

The construction project is being extended westward around Turtle Hill and is gradually encasing Montauk Point in rock. Vast quantities of glacial sediment have been removed to make way for the retaining wall, which has already covered over the strata that form the

a) Lower Till Rip Ups- Dark Streaks

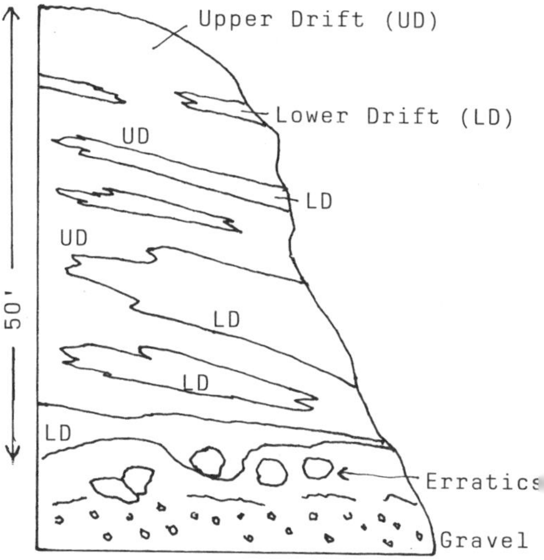


b) Lower Till Rip Ups (LD) in Upper Drift Outwash

Figure 29. Geology of Montauk Point: Views and a Geologic Column

c) Montauk Point Before Sea Wall

d) Montauk Point with Sea Wall

Figure 29. Geology of Montauk Point: Views

e) Deltaic Beds over Upper Drift

f) Sag and Sag Pond

Figure 29. Geology of Montauk Point . Continued

g) Detail of Lower Drift Rip-ups and Upper Drift Outwash

h) Boulder Layer on Lower Drift

Figure 29. Geology of Montauk Point: Views

fifty-foot hill. These layers of glacial sediment once exhibited the complex relationship of the two drifts: the lower drift, Montauk Till at the base, overlain by upper drift outwash containing ripped-up and rafted slabs of Montauk Till, and topped by a thin, upper drift till (Figure 29). The low bluff that remains to the west shows only about fifteeen feet of section with Montauk Till capped by outwash, but at the contact between the two drifts there is a one-boulder thick lag deposit at the base of the outwash. Again, a thin till caps the outwash. At the west end of the low bluff is an access road and a trail through the shrubs that leads back to the main road and the parking lot. The fifty-foot high bluffs along the beach to the west may be worth examining. They contain varying thicknesses of lower drift sediments, mainly Montauk Till, overlain by upper drift outwash with Montauk Till clasts, lake beds, and till or layers of till stones at the top.

<u>0.1 miles.</u> Leave Montauk Point parking lot and proceed westward enroute to Stop 2. The road veers steadily southwestward across the Montauk Peninsula toward the Village of Montauk.

<u>2.5 miles.</u> To the north (right) Prospect Hill rises to 160 feet in the northern upland.

<u>2.8 miles.</u> East Lake Drive. A possible side trip north to Block Island Sound with Lake Montauk on the west side and the Prospect Hill segment of the recessional moraine on the east.

<u>5.0 miles.</u> The hummocky topography of the moraine can be seen to the north in Montauk Downs Park.

<u>5.5 miles.</u> The intersection of Edgemere Avenue in Montauk village.

<u>5.8 miles.</u> The road is on a built-up causeway on top of beach deposits that closed the Fort Pond meltwater channel on the south side. The north end of the Fort Pond channel was closed

by a tombolo which now also supports a causeway for the industrial roads and the railroad.

<u>7.0 miles.</u> Montauk Sanitary Landfill. An optional stop. A large kame formed the southeast wall of the landfill. The sandy layers of the kame are deformed and laced with high angle faults resulting from either ice shove or collapse after supporting ice melted. The overlying till is relatively undeformed indicating overriding of this deposit by the glacier followed by deposition of the till as the glacier receded. The kame is an example of the recessional deposits, and its structures indicate the proximity of the ice front.

<u>Stop 2. 9.2 miles.</u> Hither Hills Overlook (Gardiners Island East Quadrangle). The overlook offers excellent views of the bays and headlands of the moraines to the west and north and of the ocean front to the southwest. The Amagansett Moraine forms the skyline to the west. In the middle distance is the Napeague Isthmus where coastal currents have deposited sand to form a tombolo, thereby filling in one of the main meltwater channels that drained the proglacial lakes to the north around twelve thousand years ago. The primary dune line adjacent to and just north of the present beach is also visible.

<u>10.7 miles.</u> Turn right (north) on East Shore Road, Napeague Harbor. Cross the railroad tracks (the morainal sediments can be seen on the right side (east) in the railroad cut). Follow the road to the end near the bay.

<u>Stop 3. 11.4 miles.</u> The Hither Hills "walking dunes" (Gardiners Island East Quadrangle, Figure 30). The entrance is on the right near the park sign. Follow the path into the woods around the west side horn of the dune, then walk up the steep dune slope to the top. Along the way there are a variety of shrubs and partially buried trees. From the rim of the dune, note

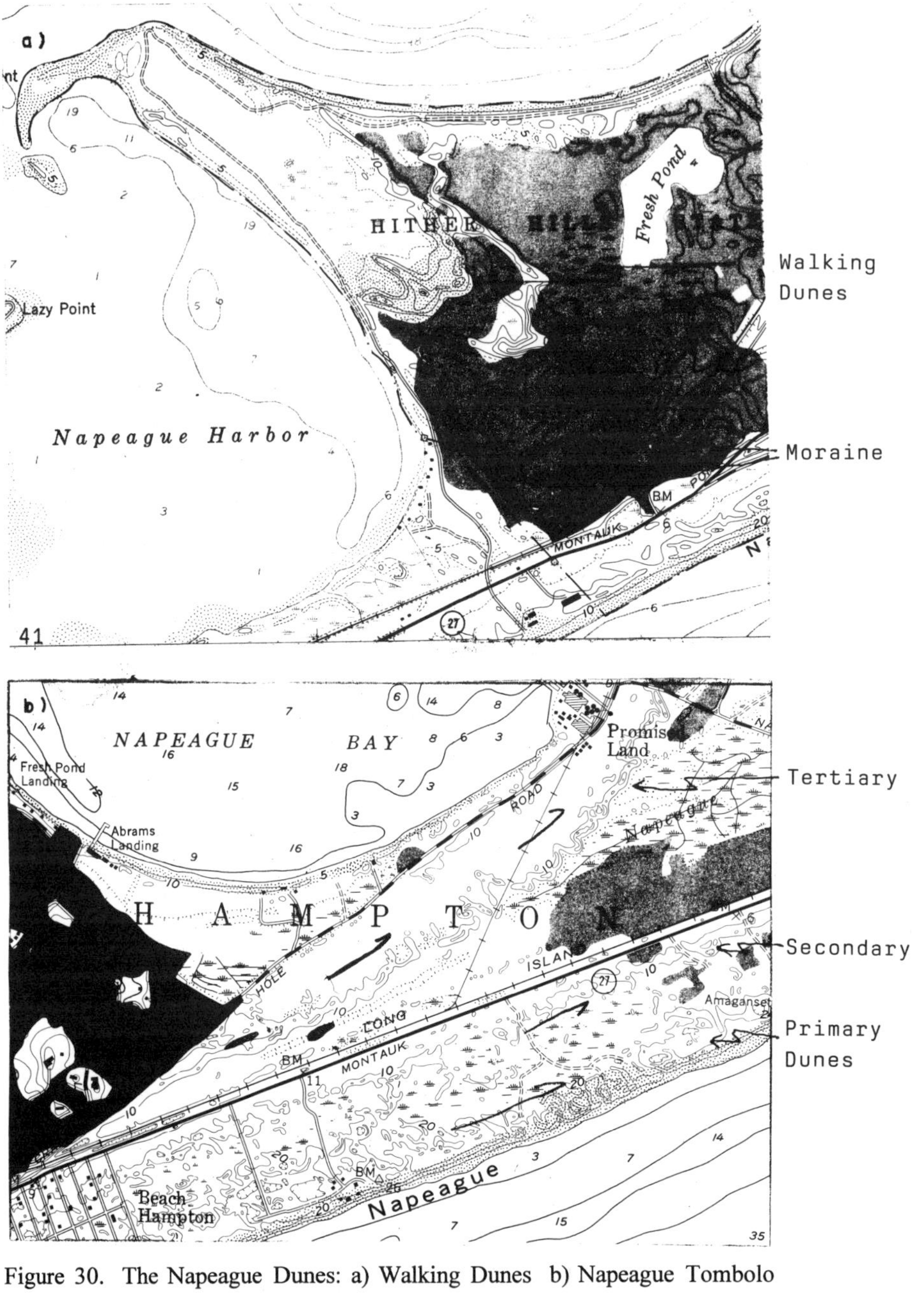


Figure 30. The Napeague Dunes: a) Walking Dunes b) Napeague Tombolo

c)

d)

Figure 30 continued. c) The Walking Dunes d) A Blowout

the following features:

1. To the southeast, the sands of a somewhat older dune can be seen banked against the moraine in the distance. This dune migrated to the southeast probably while the succeeding dunes were forming.

2. Also to the southeast, note a second dune between the one on which you stand and the dune against the moraine. Note the tree stumps that have been uncovered in the flat area as the dune migrated past. On the map, the horns of these dunes point to the north and northwest indicating that these are blowout dunes, also known as parabolic dunes, in which the horns wrap around the deflated, or blowout, area in the center and in the windward direction.

3. The forty-foot high dune on which you are standing typifies the blowout dune form with a central deflated area wrapped by the horns of the dune. A wetland occupies the central flat area of the blowout and reveals the presence of the water table. Fluctuations in the water table may have an effect on the stability of the dunes.

4. The direction of the strongest winds that actually drive the sand migration are northwest to southeast-even though the prevailing winds are southwesterly. Note the presence of smaller dune-like accumulations of sand to the north. These sand bodies may eventually coalesce to form a larger dune that will continue to migrate to the southeast. The northwesterly lowland and the Goff Point sand bar are the source areas for the dune sand. These beaches have been fed by sand derived from the bay currents.

5. The mutiple dune development and continuing sand accumulation show that these eolian processes have persisted through much of Holocene time. Note on the map that on the west side of Napeague Harbor at Napeague there is a similar dune formed southeast of Lazy Point

Road.

Return to the harbor road and proceed to Rt 27.

12.1 miles. Turn right (west) on Rt 27. For the next four miles, between the east side of Napeague Harbor and Beach Hampton, the highway crosses the Napeague Isthmus (Gardiners Island East and Napeague Beach quadrangles). Several alignments of dunes can be seen along the tombolo: first, the primary dune line adjacent and parallel to Napeague Beach; a secondary dune line runs north of and parallel to the primary dunes, just south of Montauk Highway. Most of the dunes are aligned in a longitudinal dune pattern, southwest to northeast due to the prevailing winds, but with northwest-southeast blowouts and horns due to the strong northwest wind (Figure 30).

16.1 miles. The highway leaves the sandy barrier beach, skirts the escarpment in the Amagansett Moraine north of Beach Hampton, and within one half mile enters the outwash plain. A line of dunes tails off the southeast edge of the moraine and runs northeastward toward Promised Land. The dunes in this line have a prominent northwest-southeast blowout and horn development.

17.4 miles. Rt 27 enters the outwash plain, which consists of a wedge of outwash sand and gravel that forms a triangle over ten miles wide between Amagansett and Southampton.

Stop 4. 18.4 miles. South of Amagansett, the terminal moraine is eroded with a bluff facing south toward Beach Hampton and the sea. This erosion may have taken place when sea level reached the moraine and before the beach plain was deposited. Just north of Amagansett, the terminal moraine trends southeast-northwest toward Three Mile Harbor. It then trends more east-west to just south of Sag Harbor (Sag Harbor Quadrangle). The Amagansett segment has kamic hills, mainly up to 150 ft high, but with one hill 190 ft high.

21.4 miles. Junction of Rt 27 with Rt 114, Sag Harbor-East Hampton Turnpike. Rt 114 traverses the outwash and cuts from southeast to northwest through the moraine and then into a low area of outwash southeast of Sag Harbor.

26.9 miles. Junction with Sagg Road (the starting point of Field Trip 2).

Stop 5. 27.2 miles. Cross the stream channel that links Poxabogue Pond, just beyond the trees to the north, and Sagaponack Pond to the south. This stream occupies the southern reach of a late glacial meltwater channel that formed along the axis of the interlobate angle between the Amagansett segment and the Southampton segment of the terminal moraine when the glacier occupied a recessional position to the north. Meltwater flowed from the ice southward through the Sag Harbor area, breached the moraine, and followed the southeast chain of the ponds, or pater noster lakes, along this trend from Round Pond in the north to Long Pond, Crooked Pond, Black Pond, and into Poxabogue Pond and Sagaponack Pond (Figure 31). The northerly ponds, like Round Pond and Long Pond, in the moraine, probably formed as kettles from ice blocks stranded as the glacier melted. The Long Pond segment of the meltwater channel is only about ten feet above sea level and is dammed to the south. An outwash fan to the south raises the elevation of the channel to about twenty feet at Crooked Pond.

The southerly ponds from Crooked Pond south lie in the meltwater stream channel on the outwash plain. These ponds probably originated as kettles in pitted outwash near the moraine where ice blocks were rafted into the channel and by melting of permafrost in the outwash beneath the channel to the south. The channel appears to widen south of Crooked Pond and meander to the southeast and then south between Black Pond, Poxabogue Pond and

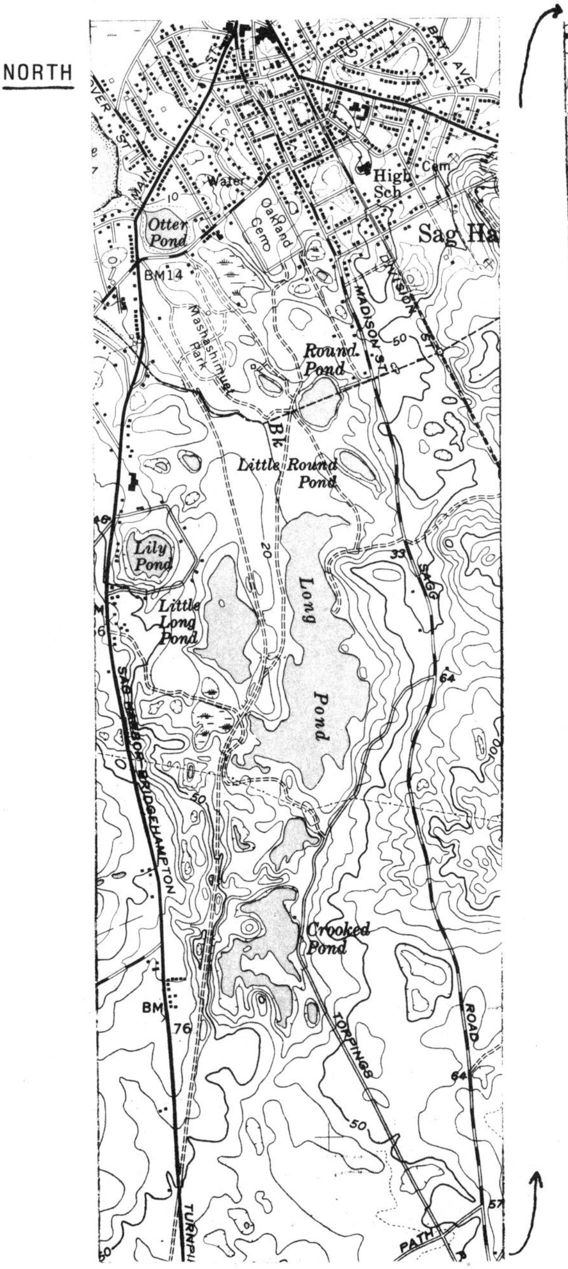


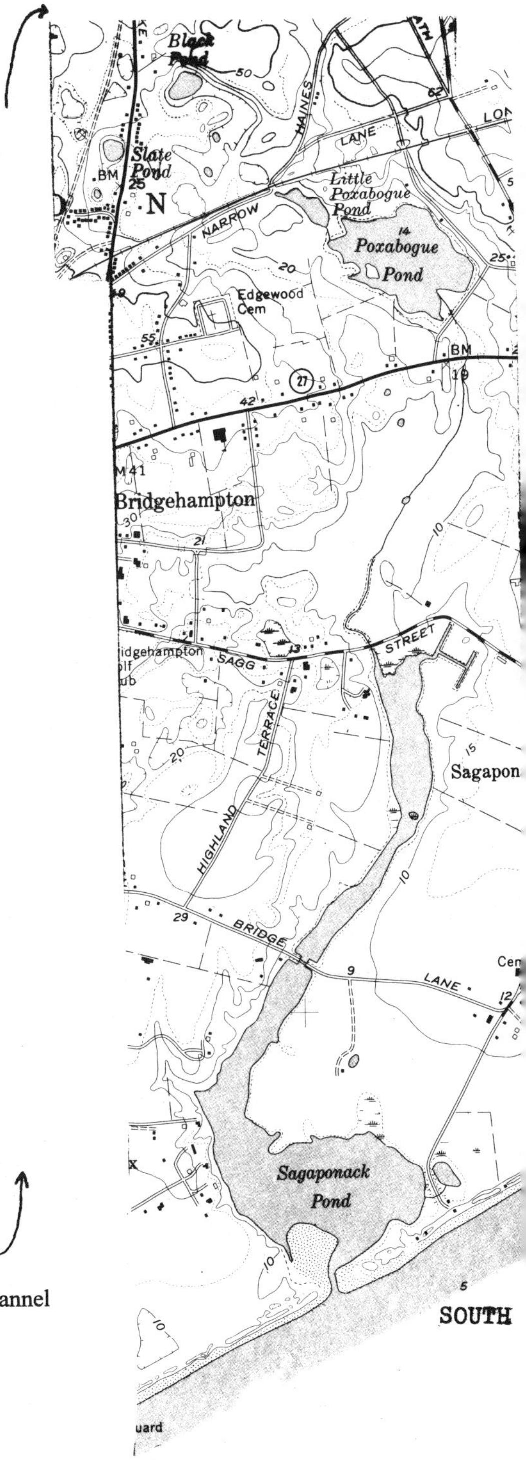


Figure 31. The Long Pond-Sagaponack Pond Channel

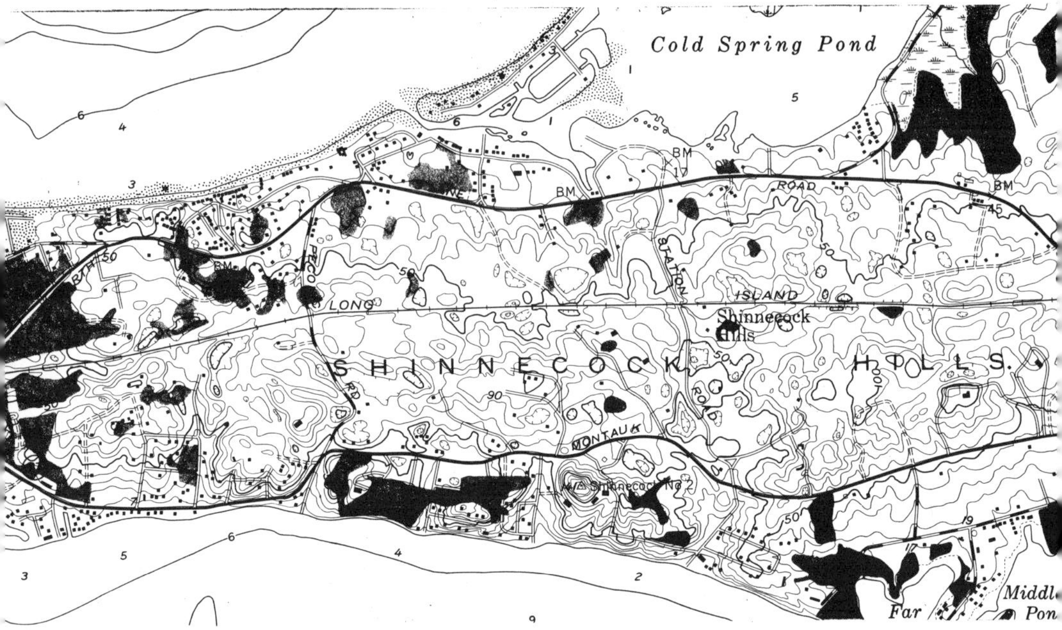


a)

b)

FIGURE 31 Continued. a) Map of Shinnecock Hills Moraine
b) Excavated Kame

Sagaponack Pond. Poxabogue Pond may have been dammed for a reservoir, while Sagaponack Pond, an actual segment of the stream valley, has been drowned by the rising sea and is now a tidal estuary connected to the Atlantic Ocean.

West of the Sag Harbor channels, the terminal moraine trends northeast-southwest from the Sag Harbor Quadrangle into the Southampton Quadrangle. This change in trend is the result of the intersection of two sublobes of the ice front which formed an interlobate angle. The easterly, Amagansett, segment of the moraine derived from the northeast, and the westerly, Southampton, segment from the northwest. The Southampton segment is significantly wider south to north and higher with summits in the two hundred to two hundred fifty foot range.

28.2 miles. Intersection of Rt 27 and Sag Harbor-Bridgehampton Turnpike. Northwest of Bridgehampton, the Southampton segment of the terminal moraine is conspicuously higher as seen from the highway.

29.3 miles. The Scuttlehole Ponds (Sag Harbor Quadrangle). The highway crosses another meltwater channel. This one heads in the moraine nearly two and one half miles to the northwest. This channel also has a chain of pater noster lakes aligned in a northwest-southeast trend between Shorts Pond, Haines Pond, Goldfish Pond, Long Pond and Little Pond, to Kellis Pond. At this point an outlet drains Kellis Pond southwest into Calf Creek and Mecox Bay suggesting a recent modification of the stream valley. The original course may have been into Swan Creek. The inlet into Mecox Bay has also been periodically dredged. The Scuttlehole channel drained meltwater from the terminal moraine southward, but did not extend to the proglacial lakes north of the moraine. The northern kettles through the north end of Long Pond may have formed in ice block depressions. The southerly ponds may be dammed reaches of the former stream valley.

31.8 miles. Water Mill. Cross the dammed and modified Mill Pond-Mill Creek-Mecox Bay drainage.

32.8 miles. Bear right on Rt 27.

34.8 miles. Intersection with North Sea Road (Southampton Quadrangle).

35.5 miles. Intersection with Tuckahoe Lane. West of this point the highway traverses the terminal moraine.

35.5 to 40.0 miles. Southwest of the Big Fresh Pond-Little Fresh Pond meltwater channel, north of Southampton, the terminal moraine swings westward into the Shinnecock Hills region. This may represent another interlobate angle in the moraine. It does signify a marked change in the style of the moraine. The Shinnecock Hills segment of the moraine forms a narrow east-west band of kamic hills that range between fifty and one hundred feet high. Many of the hills along the newer section of Rt 27 have been exploited for sand and gravel, and the mined areas are now occupied by residential and commercial structures (Figure 31).

Stop 6. 35.5 miles. Northwest side of Tuckahoe Lane intersection, adjacent to the shopping plaza. Stop here to view a remnant of one of the kames in which the strata, typically a thin stony till overlying sandy outwash, can be seen.

**FIELD TRIP 2.**

A south to north traverse from the outwash plain through the terminal moraine and the recessional moraines of eastern Long Island (Figure 32; Sag Harbor and Greenport Quadrangles).

<u>0.0 miles.</u> This trip begins at the junction of Montauk Highway, Rt 27, and Sagg Road, east of Bridgehampton, at approximately mile 27 of Field Trip 1. Proceed north on Sagg Road crossing the outwash plain east of Poxabogue Pond. Sagg Road runs parallel to the Long Pond meltwater channel and chain of lakes.

<u>1.7 miles.</u> Cross an area of pitted outwash and loess-rich soil.

<u>Stop 1. 1.8 miles.</u> This is the approximate contact between the outwash plain and the distal slope of the terminal moraine, east of Crooked Pond, which appears to straddle the contact.

<u>2.4 miles.</u> West of the southern half of Long Pond is the site of the old Town landfill, which can be entered from Sag Harbor-Bridgehampton Turnpike. The landfill pit was dug into morainal sediments and revealed till of the lower drift at the base of the excavation.

<u>Stop 2. 3.0 miles.</u> Kame and kettle topography is prominent, making up the morainal hills east of Little Round Pond. To the west is the modern northwest-trending valley in a possible meander scar left over from the original south-flowing meltwater channel. It is now occupied by Ligonee Brook, an underfit stream flowing from Long Pond at about ten feet above sea level to Sag Harbor Cove. The village of Sag Harbor and the adjacent Sag Harbor Bay occupy the lowland to the north. Sagg Road merges into Madison Street.

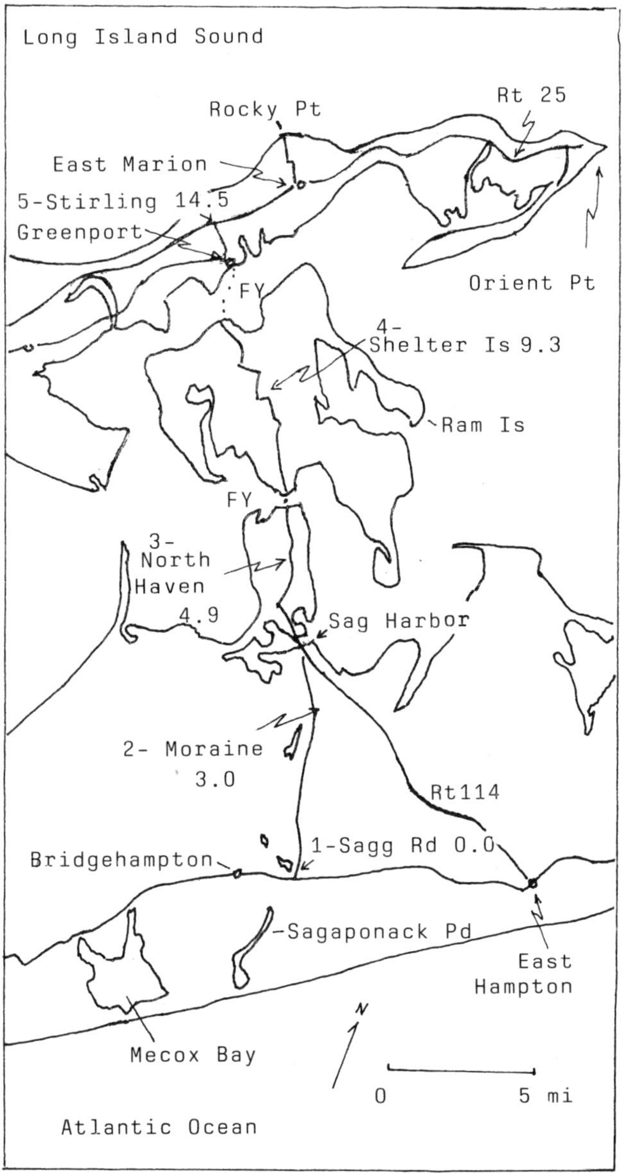


Figure 32. Route of Field Trip 2

4.2 miles. Junction of Madison with Main Street. Continue northward. Sag Harbor is on a low plain underlain by lower drift till. Stream valley alluvium fills the valley, and marine sediments and marsh deposits cover the coastal areas. Along the northern margin of the village, Sag Harbor Cove, southern Noyack Bay to the west, and Sag Harbor Bay to the east are relicts of east-west drainage along the margin of the glacier north of the terminal moraine at the Sag Harbor position (Figure 33).

4.5 miles. Junction with Rt 114 (Greenport Quadrangle). Bear left over the bridge to North Haven. Rt 114 leads across North Haven and Shelter Island through a number of turns. Stay on Rt 114.

Stop 3. 4.9 miles. Rt 114 rises onto the generally low-lying North Haven terrain. North Haven is an island tied to the mainland of the South Fork by tombolos, although the eastern sand bar is now submerged. The peninsula is underlain by lower drift till, and its surface is covered with a veneer of upper drift ground moraine and kames that range from thirty to eighty feet high. The kames are concentrated along the east and west sides of the peninsula and contain abundant till stones and erratics. On the west shore, the kame on which the Retreat House stands is eighty feet high. It is made mostly of outwash, capped by a few feet of sandy and stony till. These upper drift kames and surficial sediments show that North Haven is a segment of a recessional moraine of the last glaciation. In the reconstruction of recessional moraines, North Haven occupies the interlobate position between a western lineation of related deposits from Cow Neck to Jessup Neck to North Haven that are associated with the Connecticut Lobe of the last glacier, and the eastern lineation from North Haven to Fireplace with the Connecticut-Rhode Island sublobe of the glacier.

a)

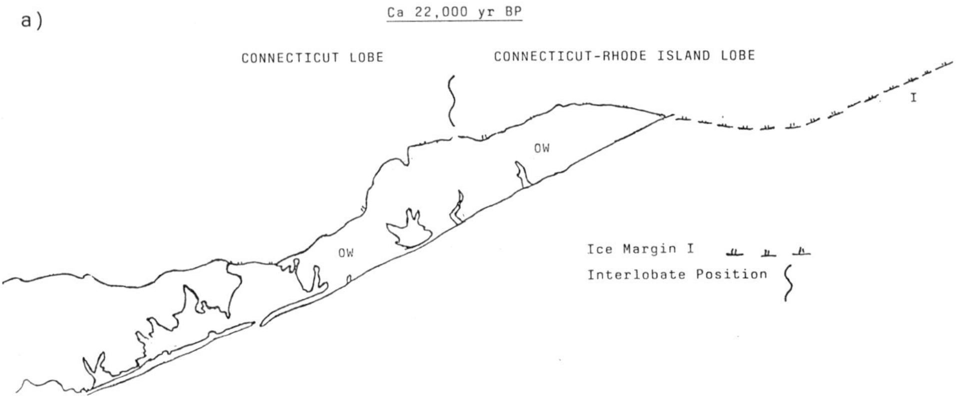


a) Ice Margin I. Late Wisconsinan Maximum, Terminal Moraine
b) Ice Margin II. Late Wisconsinan Recessional Moraine
c) Ice Margin III. Late Wisconsinan Recessional Moraine
d) Ice Margin IV. Late Wisconsinan Recessional Moraine
e) Long Island ca 20,000 BP. Ice Free
f) Draining of Proglacial Lakes ca 12,000 BP

( in e and f: MW= meltwater channel )

Figure 33. Late Wisconsinan Ice Margins

b)

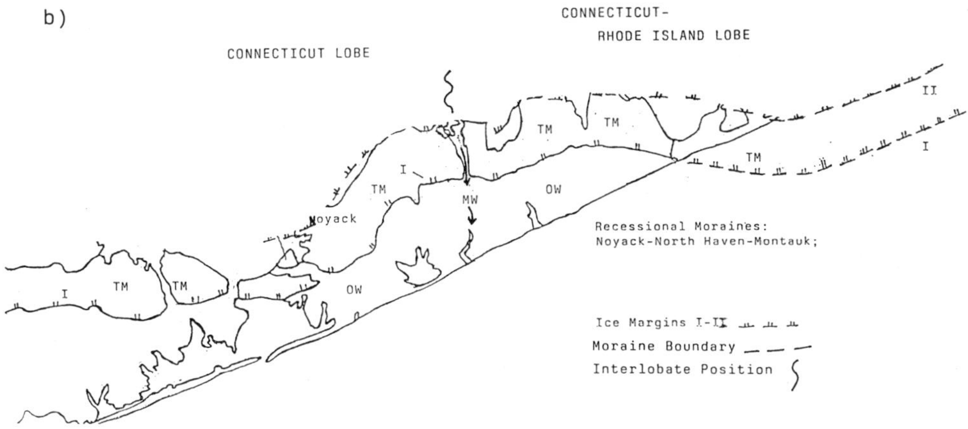


c)

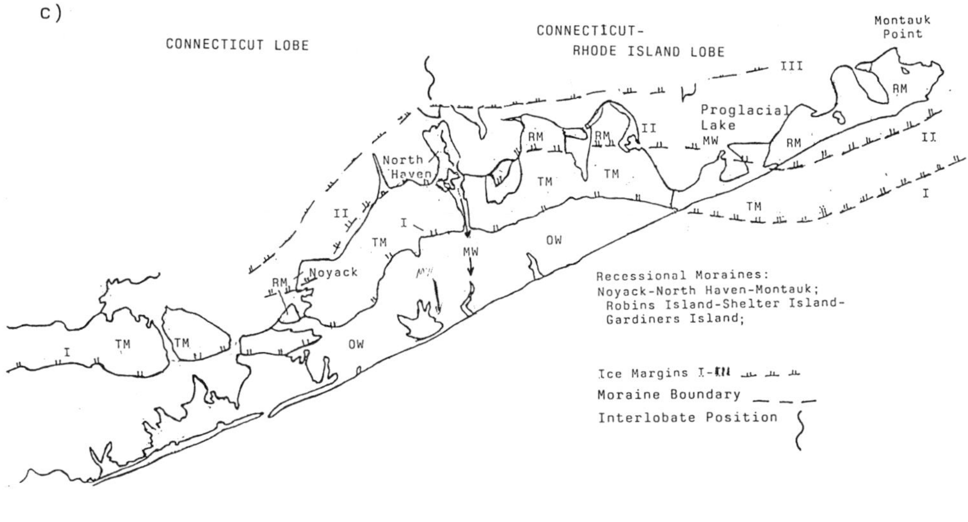


Figure 33. Late Wisconsinan Ice Margins Continued

d)

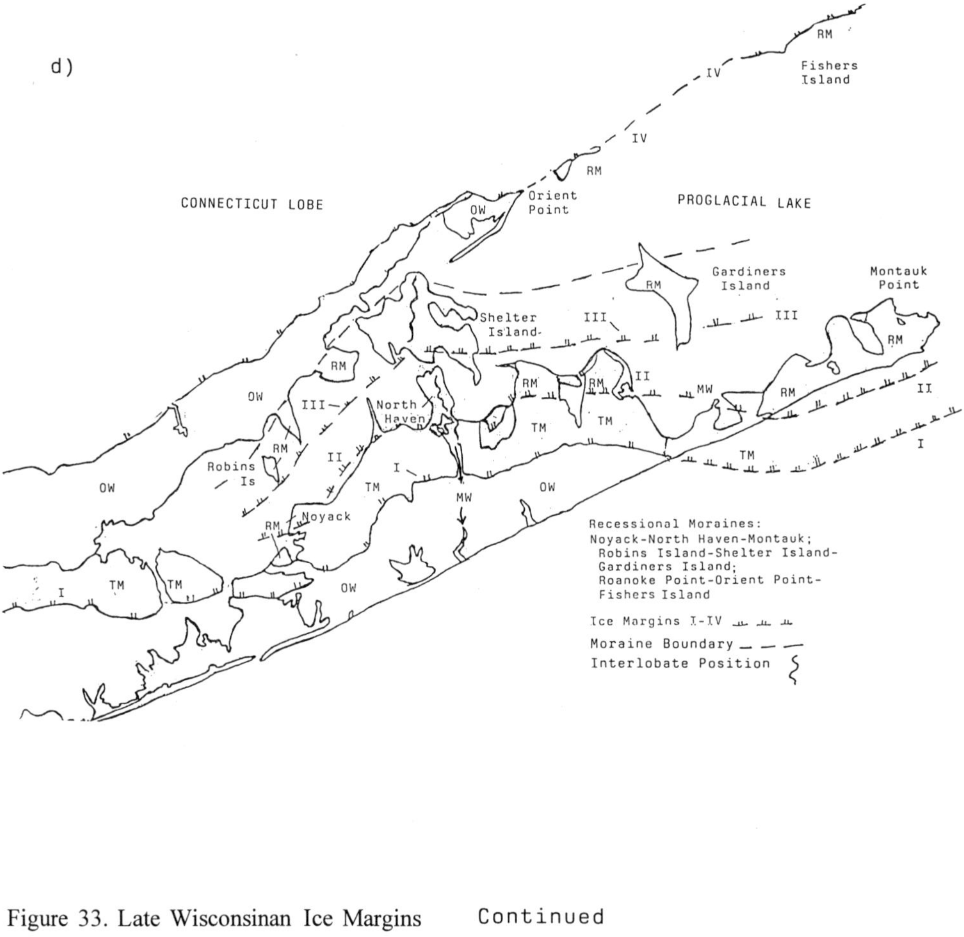


Figure 33. Late Wisconsinan Ice Margins Continued

e)

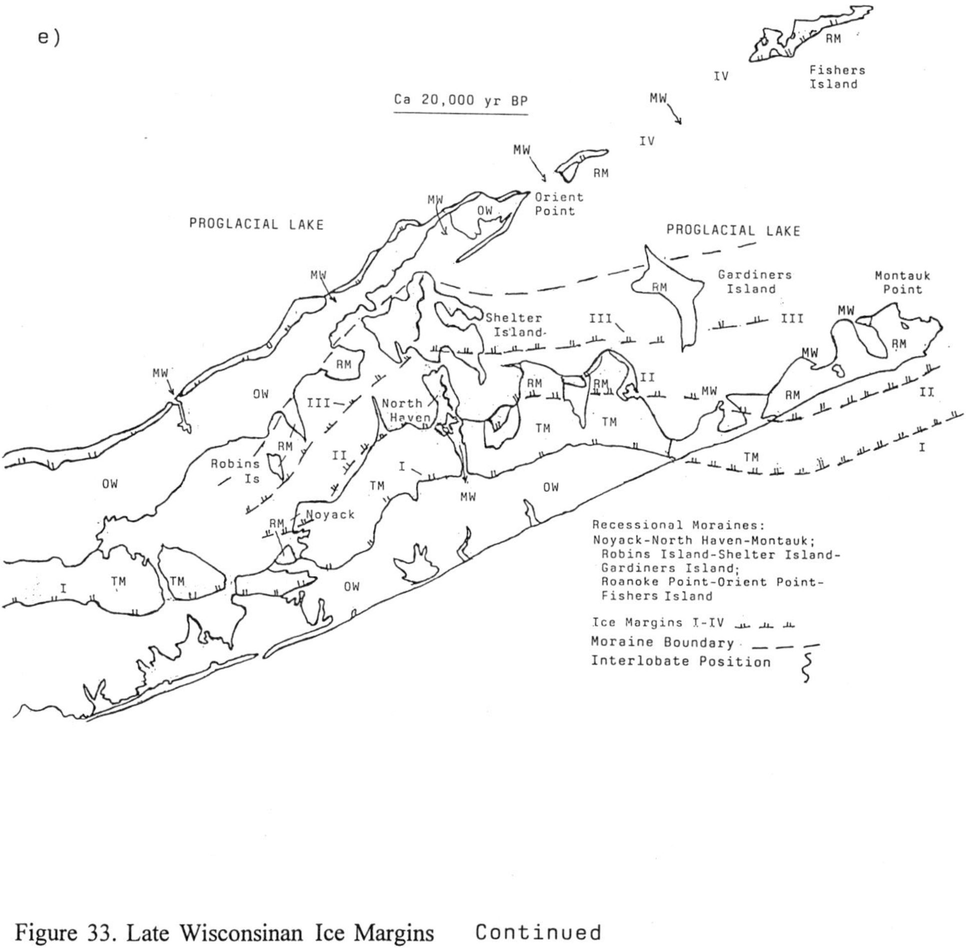


Figure 33. Late Wisconsinan Ice Margins Continued

f)

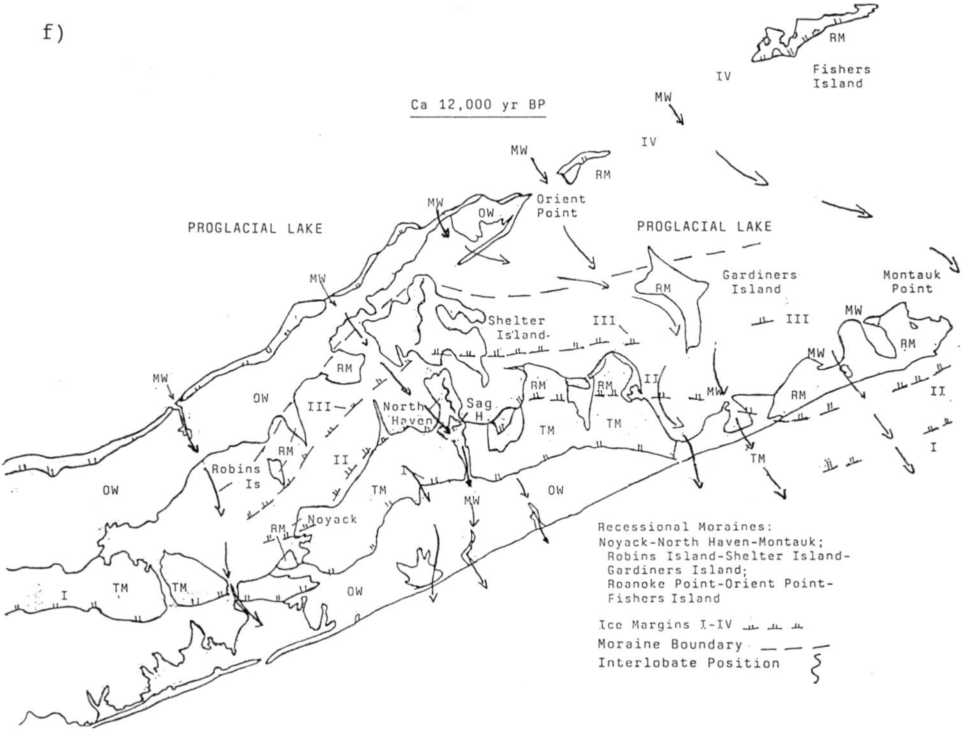


Figure 33. Late Wisconsinan Ice Margins Continued

5.7 miles. Turn right (north) on Rt 114. The road traverses low areas where drainage is to the east and reaches a drainage divide at about 6.5 miles.

6.7 miles. Northward from the divide, the drainage is to the north, mainly via a small stream into Shelter Island Sound. The road cuts through a kame that forms the northeastern section of the peninsula.

7.7 miles. Proceed to South Ferry landing. Take ferry across Shelter Island Sound to Shelter Island. Note bluffs in moraine on Shelter Island to the northeast.

8.1 miles. Shelter Island occupies the interlobate position along the Robins Island-Little Hog Neck-Great Hog Neck-Shelter Island-Gardiners Island Recessional Moraine and ice margin.

Stop 4. 9.3 miles. Junction with Manhasset Road. Turn left (west) on Rt 114. The alignment of morainal hills to the east and southeast represents the east-trending segment of the recessional moraine (Figure 32). The range of hills across northern Shelter Island represents the western segment of the recessional moraine. The two separate morainal segments are separated on Shelter Island by a low-lying central area, roughly parallel to Smith Road. The offset of the trends of the two morainal segments in the interlobate zone indicates that they, in fact, overlap, as did the glacial lobes that formed them.

9.5 miles. Rt 114 bears right.

9.8 miles. Turn left (west) on Ferry Road (Rt 114).

10.2 miles. Turn right (north) on Ferry Road (Rt 114) through Shelter Island village and into the hummocky northern terrain. The kames forming Shelter Island Heights have elevations above one hundred eighty feet, with significant exposures of the morainal sediments in the bluffs along the northwest coast.

Optional Side Trip. Take Winthrop Road east to Ram Island Road following road along beaches with views of headlands in the moraine on Ram Island and Little Ram Island.

11.0 miles. Left (west) on West Ferry Road.

11.5 miles. Right (north) on North Ferry Road (Rt 114).

12.0 miles. The road turns left through village, crosses north-flowing drainage at Chase Creek, and then skirts the edge of Shelter Island Heights.

12.5 miles. Proceed to the North Ferry landing. Take the ferry across the harbor to Greenport. Enroute observe the bluff in the moraine to the south.

13.5 miles. Greenport. Although much of the surface is developed, the village overlies a low-lying outwash plain that extends from the Roanoke Point Moraine to the north. Turn right on Front Street, left on Main Street and proceed northward to Stirling.

Stop 5. 14.5 miles. Stirling. Intersection with Rt 48 and Field Trip 3. The outwash fan originates north of Stirling in the narrow remnant of the Roanoke Point Moraine. The moraine forms low bluffs along the north coast and can be seen at Stop 3 of Field Trip 3 (Figure 33).

**FIELD TRIP 3.**

The recessional moraine and outwash plain of the North Fork and north shore of eastern Long Island (Figure 34).

Orient Point and Plum Island to the east are segments of the Roanoke Point-Orient Point-Fishers Island recessional moraine of the Connecticut Lobe and Connecticut-Rhode Island sublobe of the Laurentide glacier. The Orient Point segment meets the Roanoke Point segment in an interlobate angle-where the two glacial lobes converged-at East Marion. The Fishers Island segment continues eastward toward the Rhode Island coast where it joins the Charlestown segment of the recessional moraine. The lobate nature of the recessional moraines is confirmed by the till fabric, which is determined by the orientation of till stones, and rock types taken sequentially from upper till exposures in bluffs on the long, east to west curve of the Roanoake Point Moraine that borders Long Island Sound. Till stones show a closer affinity to bedrock sources more directly to the north; those taken between Riverhead and Wading River match the Hartford Basin rock types in the Connecticut River Valley, and the till fabrics point in that direction. Fabrics from the west end of the moraine point to the northeast; those from the east side to the northwest. This evidence gives credence to the moraine's concave shape-as inherited from the Connecticut Lobe- as opposed to an erosional origin for the arc of the moraine. The late Wisconsinan age of the moraines is corroborated, in part, by the pollen data: all of the pollen records from bogs on morainal deposits of Long Island-terminal

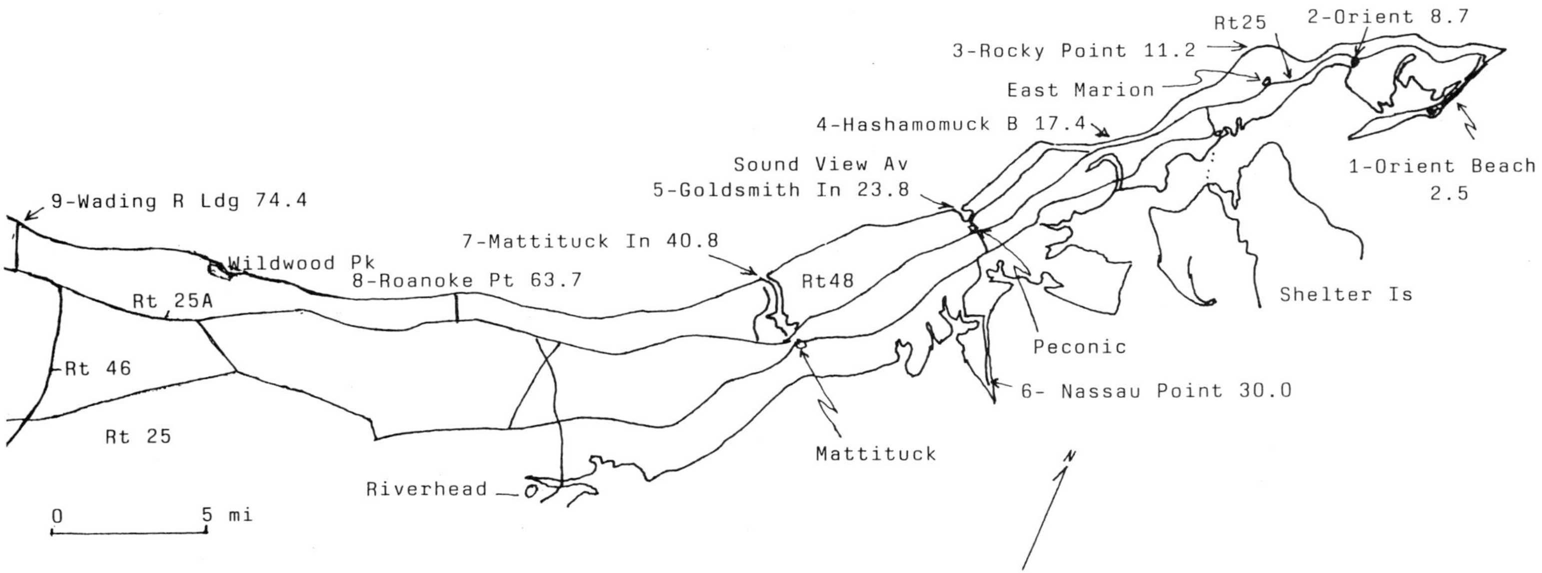


Figure 34. Field Trip 3 with Field Trip Stops and Mileage

moraine or recessional moraine- postdate the last glaciation and have the cold to warm, postglacial sequence of climate as inferred from the pollen zonation. The post and supra-terminal moraine bog sediments contain substantial evidence of the herb pollen zone, while the recessional moraine bogs have a limited herb zone record, and both are succeeded by spruce, pine and oak zones (Figure 11).

The field trip begins at Orient Beach State Park. The Park occupies an elongate sand and gravel spit, or barrier beach, that has migrated nearly five miles westward on east to west currents, enclosing Long Beach Bay and forming the southeastern margin of Orient Harbor.

<u>Stop 1. 0.0 to 2.5 miles.</u> Orient Beach State Park Entrance (Plum Island Quadrangle). Enter the Park from Route 25. Follow the road for two and one half miles to the Beach 1 Parking Lot (Orient Quadrangle). The Park road is on a causeway built above the barrier beach. Gardiners Bay is on the southeast side of the beach, Little Bay is to the northwest. Salt marsh borders the barrier on the Little Bay side. The sediment that makes up the natural beach of the barrier is coarse-grained sand and fine quartz gravel derived directly from the outwash of the moraine at Orient Point and carried westward into Gardiners Bay. Along the barrier, Plum Island, Orient Point Lighthouse and the ferry terminal can be seen to the northeast, Gardiners Island to the southeast, and Shelter Island to the southwest. The erratics used in the landscaping around the beach were probably hauled in from the moraine (Figure 35). Trees and shrubs on the barrier include scrub oaks, junipers, old field cedars, azaleas and occasional pines.

<u>5.0 miles.</u> Turn left (west) on Rt 25. The road crosses the outwash plain south of the distal slope of the recessional moraine. The outwash was deposited by the receding glacier between

a) Goldsmith Inlet. West-trending spit on right. East-trending spit at surf line.

b) Roanoke Point Moraine. Note erratics; residual gravel berm.

Figure 35. Views of the North Shore

the Roanoke Point-Orient Point ice margin and the Robins Island-Shelter Island-Gardiners Island Recessional Moraine to the south. Between Mulford Point and Orient Point, the Orient Point Moraine segment has been eroded away. Pettys Bight occupies a deep erosion scar along the shoreline where the moraine and substantial outwash have been removed (Orient Quadrangle). West of Mulford Point, remnants of the moraine form Browns Hills and Terry Hill, but the moraine is missing where a meltwater channel breached the moraine west of Terry Point.

8.5 miles. North of Orient, there is a view of the salt marsh and bay to the south.

Stop 2. 8.7 miles. Stop where convenient. The glacial meltwater channel cut through the moraine between Terry Point and west of Dam Pond. Truman Beach on the north shore and the sand bars on the south side, have formed a double tombolo which closed the channel and reconnected the Orient Point and East Marion segments of the North Fork in more recent time (Figure 36). The meltwater channel drained the proglacial lake to the north and crossed Gardiners Bay as a tributary of the main channel that trimmed the Amagansett Moraine (the terminal moraine) at Acabonack and cut through the recessional moraine at Napeague Beach.

9.2 miles. An optional stop pending access to the parking lot. Orient Point Bathing Beach and Long Island Sound. To the east, large erratics eroded from the moraine armor the beach at Terry Point. Similarly, erratics lie on the beach next to the moraine to the west toward Rocky Point. The beach is made of coarse brown sand and small-size quartz pebble gravel from the outwash. Proceed west on Rt 25. Dam Pond is on the right (north), Orient Harbor on the left (south).

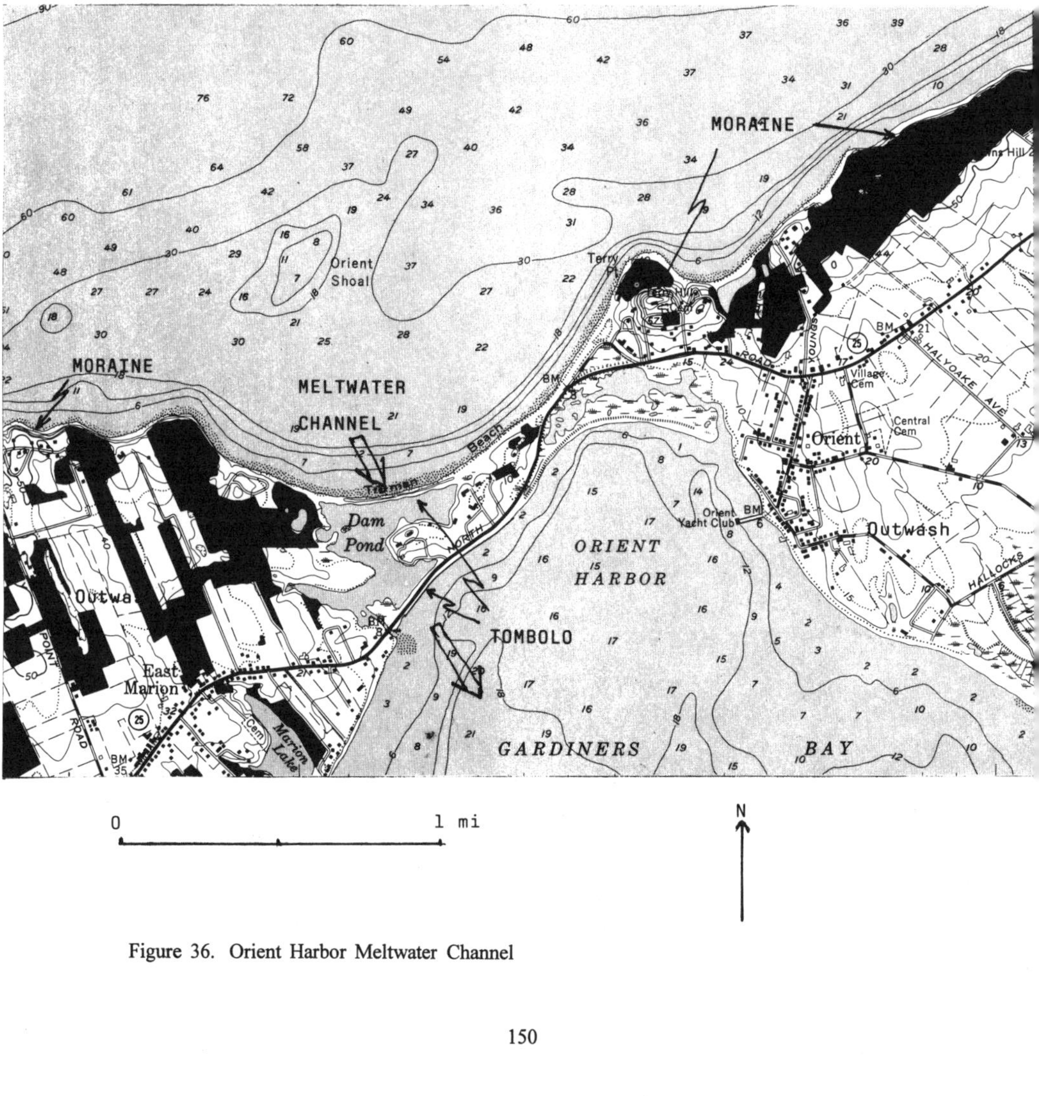


Figure 36. Orient Harbor Meltwater Channel

9.5 miles. Enter the village of East Marion where the road ascends the moraine. The apparent change in direction of the segments of the recessional moraine at East Marion (the Orient Point segment trends more east-west; the Roanoke Point segment northeast-southwest) describes the interlobate angle between lobes of the last glacier.

10.0 miles. Turn right (north) on Rocky Point Road, crossing an outwash fan, toward the moraine and Long Island Sound.

11.0 miles. Large erratics appear on the right side, marking the transition from the outwash fan to the moraine.

Stop 3. 11.2 miles. At Rocky Point, a fifty foot exposure of glacial strata consists of nearly ten feet of bouldery till overlying the coarse sand and fine quartz pebble gravel of the outwash, and is characteristic of the Roanoke Point Moraine.

12.4 miles. Return to Rt 25 in East Marion. Proceed west on Rt 25 across the outwash plain to Stirling (Greenport Quadrangle).

15.1 miles. Rt 25 merges into Rt 48, North Road.

15.6 miles. Stirling. Intersection with Field Trip 2. Stay on Rt 48. Rt 25 turns south to Greenport.

17.2 miles. Rt 48 traverses small outwash fans. North of Chapel Lane, a remnant of the Roanoke Point Moraine forms a headland.

Stop 4. 17.4 miles. Stop where convenient. Hashamomuck Beach (at the motel complex) (Southold Quadrangle). Hashamomuck Pond overlies a meltwater channel that eroded through the moraine here (Figure 37). The channel and its tributaries also include the marsh north of Arshamonaque. The road follows a tombolo across the north side of the channel; the southerly side is also sealed by a tombolo. Note the sand bars at Mill Creek. Meltwater once flowed from the proglacial lake across Shelter Island Sound and into the main channel to the east.

17.9 miles. Southold Town Beach. Sand seems to be more abundant than gravel on this section of beach.

18.9 miles. Bear right (west) at the entrance to Sound View Avenue.

19.9 miles. The road crosses the pitted and stream dissected outwash plain before reaching the Roanoke Point Moraine at Horton Neck.

20.6 miles. The hummocky moraine forms a headland at Horton Point. Because of beach erosion, numerous groins have been built along the beach to protect homes. At Lighthouse Road, Sound View Avenue bisects a small kettle pond in the pitted outwash and then curves southwestward through a stand of maples on the moraine.

20.8 miles. At Horton Lane, oak is a dominant tree in the hardwood forest. Lily Pond is the remnant of a northwest-trending valley whose stream has been sealed off by beach deposits.

21.3 miles. Kennys Road. Sound View Avenue again traverses the moraine. Erratics are abundant on the surface here. Great Pond is also a relict of northwest-trending drainage.

22.2 miles. Pine Crest Dune Camp. The beach here is much wider due to excess sand in the longshore drift and perhaps a balance of deposition between east and west currents.

23.1 miles. Junction with the valley leading north to Goldsmith Inlet (Southold Quadrangle), a Holocene valley whose northwest-flowing stream drains the glacial upland. It was nearly cut off from Long Island Sound by a northeast to southwest oriented spit (Figure 37).

23.2 miles. Turn right (north) on Mill Lane.

23.5 miles. The road cuts through the moraine onto the beach plain.

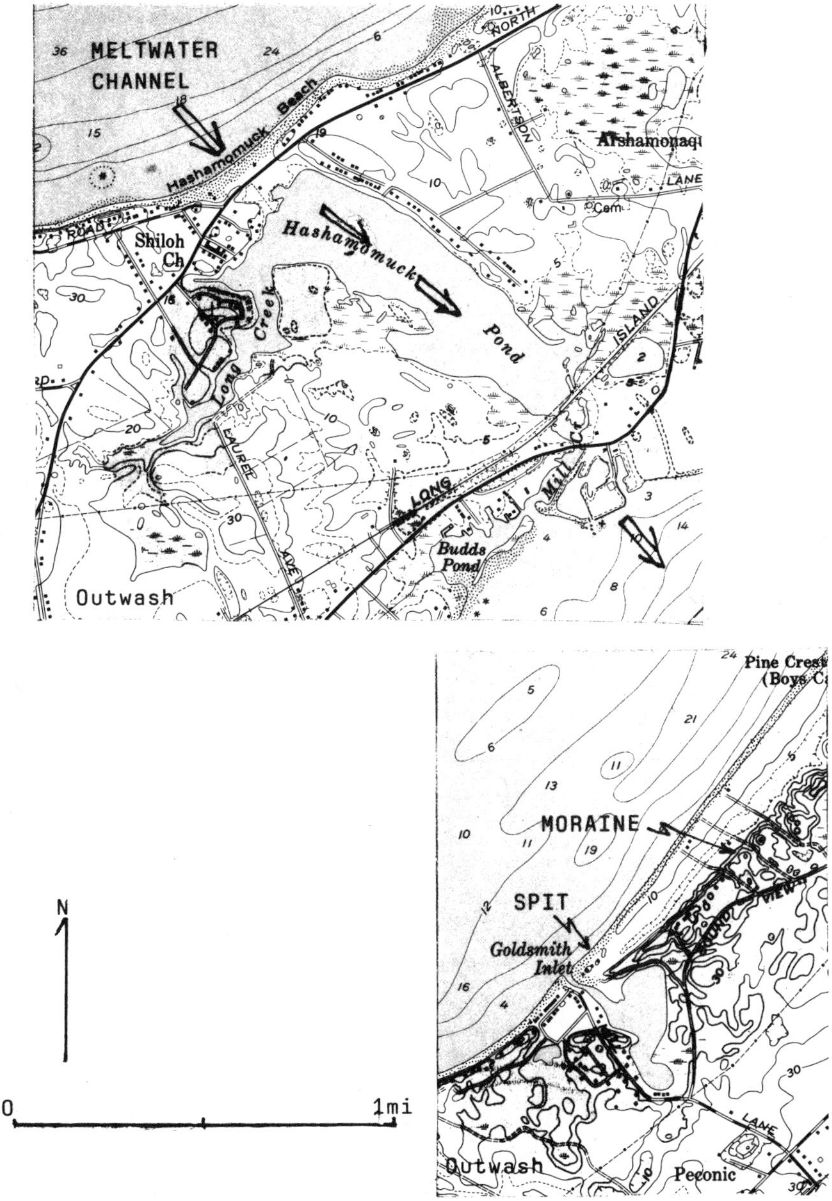


Figure 37. Hashamomuck Meltwater Channel and Goldsmith Inlet

Stop 5. 23.8 miles. Goldsmith Inlet (Southold Quadrangle). The beach sediment is mainly quartz pebble gravel. Note the effect of the jetty on sand accumulation. While the inlet has been partly closed by a northeast to southwest spit, at the present time sand is building up on the west side of the west jetty, and the beach is eroding on the east side of the east jetty. Furthermore, the channel is being deflected to the east and into the Sound. The current responsible for the west to east drift appears to be the opposite of the longshore drift direction that built the original spit. However, it may be that the west jetty has interfered with the net drift of sediment that occurred naturally and deposited the spit. Return on Mill Lane to Rt 48 at Peconic.

24.5 miles Junction Rt 48. Proceed south on Peconic Avenue for a side trip to Little Hog Neck (Southold and Southampton Quadrangles). The route crosses the intermorainal outwash plain between the Roanoke Point Moraine to the north and Little Hog Neck, which is a segment of the Robins Island-Little Hog Neck-Great Hog Neck-Shelter Island-Gardiners Island Recessional Moraine (Figures 5, 12 and 33).

25.1 miles. Junction Rt 25. Turn right (west) on Rt 25.

26.1 miles. Turn left (south) on Bay Avenue.

27.4 miles. Turn left (east) toward Broadwater Cove.

28.0 miles. Turn right (south) on Nassau Point Road. Cross the tombolo that connects Little Hog Neck to the rest of the North Fork.

Stop 6. 30.0 miles. Nassau Point (Southampton Quadrangle). The moraine at Little Hog Neck is comprised of outwash capped by brown till. It has characteristic hummocky topography with kamic hills, some over ninety feet high. The east side of the Neck has been trimmed by bay currents into fifty foot high cliffs.

33.3 miles. Return to the junction of Bay Avenue and Rt 25. Proceeding westward, cross the outwash plain which has been dissected by south-flowing streams. The outwash plain is home to farms for produce, sod and grapes.

35.0 miles. Cutchogue.

37.8 miles. Mattituck (Mattituck Quadrangle), junction of Rt 25 and Rt 48. The village sits astride a twenty foot east-west ridge of outwash that separates the southerly drainages of Deep Hole Creek, including Marratooka Lake, and James Creek, from the north-flowing drainage of Mattituck Creek (Figure 38). During deglaciation, however, the drainage flowed southward in a meltwater channel from the proglacial lake in Long Island Sound at an elevation above twenty feet, through Mattituck Creek and its tributary, Long Creek, and through the south-trending valleys. Marratooka Lake may be a relict of that drainage and/or the result of periglacial processes in which ice melted out below the outwash. The small valley from the lake to Deep Hole Creek suggests that a prior connection existed. The Long Creek valley (Mattituck Hills Quadrangle) extends northward through the Oregon Hills segment of the Roanoke Point Moraine and the adjacent outwash fan north of Waterville and could also have drained water from the proglacial lake. Mattituck Creek has cut deeply into the outwash along its northerly course and represents the dominant meltwater channel.

37.9 miles. Continue on Rt 48.

38.9 miles. Junction with Cox Neck Road. Turn right (north) on Cox Neck Road.

39.6-39.7 miles. Cross Mattituck Creek to junction with Breakwater Road (Luthers Road). Turn left (north). The road traverses the outwash plain toward the recessional moraine.

40.5 miles. Contact between the outwash and the moraine.

Stop 7. 40.8 miles. Mattituck Inlet. The jetty on the west side of the Inlet has trapped so much sand from the west that it is virtually covered (Figure 38). Sand now by-passes the jetty and forms sand bars northeast of the Inlet. This sand movement is the opposite of the east to west drift that built the spit that formed the east side of the Inlet and forced Mattituck Creek westward. This is the same type of longshore drift reversal seen at Goldsmith Inlet in which the present, net direction of longshore drift and sand transport is eastward. We can attribute this effect to the presence of the jetties which block sand movement. Both east to west and west to east currents operate seasonally and may be tidal or storm generated. The apparent drift direction is west to east with sand piling up on the west side jetty. The adjacent Mattituck Hills are the continuation of the Roanoke Point Moraine to the west. The moraine here is somewhat broader, south to north, than the Oregon Hills reach to the northeast.

Retrace the route to the junction of Cox Neck Road and Rt 48.

50.7 miles. Rt 48. Turn right (west)

51.1 miles. Rt 48 merges into Rt 27A.

55.9 miles. Northville Tank Farm (Riverhead Quadrangle). Petroleum products, storage tanks are embedded in the distal slope of the moraine and in the adjacent, outwash fans to the north.

60.0 miles. Junction with Northville Turnpike.

60.5 miles. Good view of the distal slope of a south-trending lobe of the Roanoke Point Moraine. Lobes like this one considerably widen the outcrop belt of the moraine.

63.1 miles. Junction with Park Road. Turn right (north) to Reeves Park and Roanoke Landing (Riverhead Quadrangle).

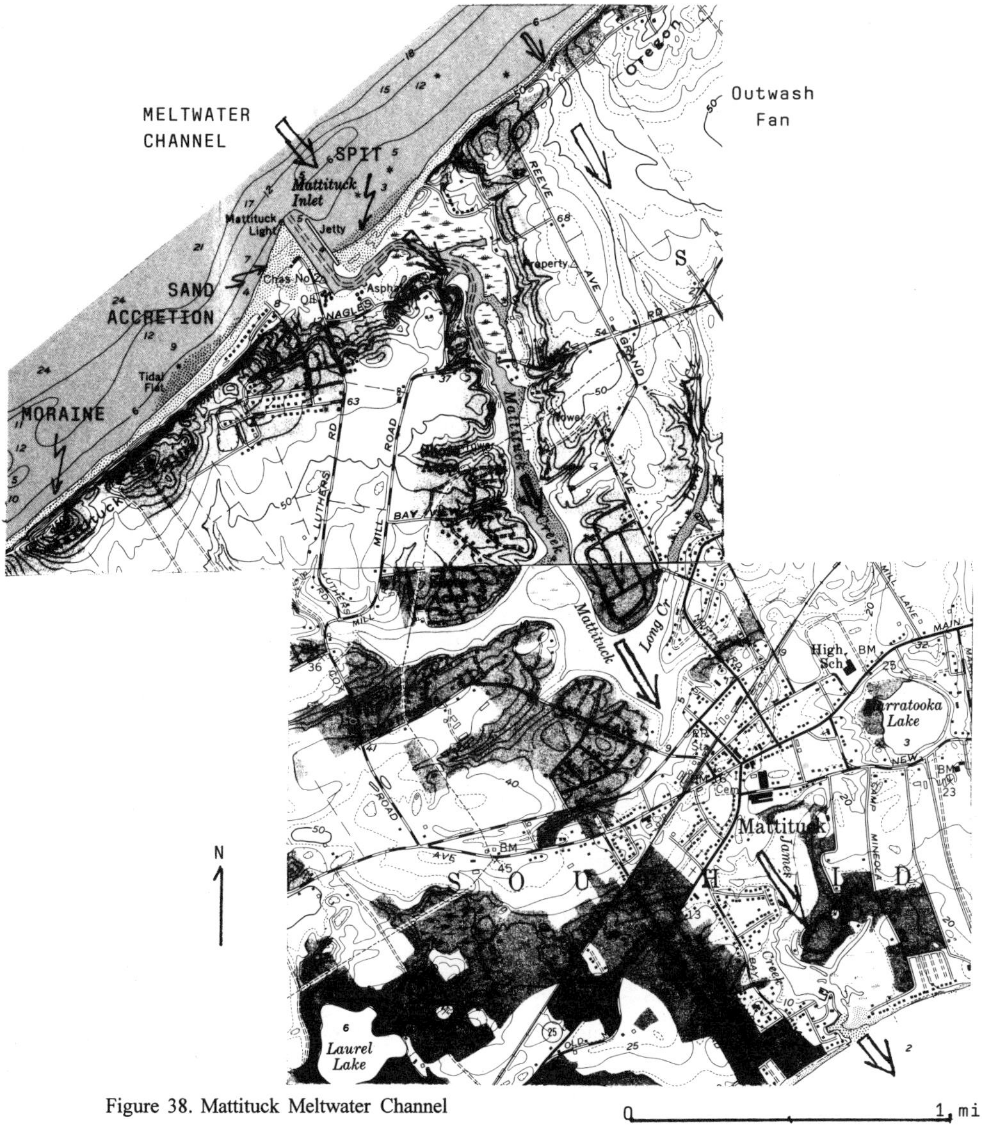


Figure 38. Mattituck Meltwater Channel

63.5 miles. The road cuts through the distal slope of another south-trending lobe of the moraine.

63.6 miles. Crest of the Roanoke Point Moraine. Descend a steep north-facing valley on the proximal slope of the moraine into the parking lot at Roanoke Landing.

Stop 8. 63.7 miles. There are good views and photo opportunities of Long Island Sound and the exposures of the morainal sediments. Recent renovation of the parking area has uncovered contorted morainal sediments on the east side of the lot. While it might appear that the contortions are ice-contact in origin, the sedimentary layers are more deformed closer to the north edge of the bluff. This indicates that the bluff-side deformation is the result of recent slumping rather than ice shove or collapse. The one hundred foot high bluff at Roanoke Landing is comprised of coarse outwash that encloses occasional masses of silt (possible proglacial lake deposits) and frequent lenses of brown flow till with ice contact fractures. The outwash is covered by two to three feet of meltout till. At the surface, about one foot of loess forms the matrix of the soil cover. The outwash consists of coarse sand and quartz pebble gravel, and the flow till is mainly a silty clay (Figure 39). Return to Sound Avenue, Rt 27A. Turn right (west) on Rt 27A.

64.3 miles. A good view to the south of the Ronkonkoma Moraine across the outwash plain and Peconic Valley lowland. Bald Hill, the high point on the moraine at 295 ft, stands out on the horizon.

66.2 miles. Baiting Hollow (Riverhead Quadrangle). The moraine has been narrowed by coastal erosion to less than 0.2 miles north to south, north of the village of Baiting Hollow on Rt 27A.

a)

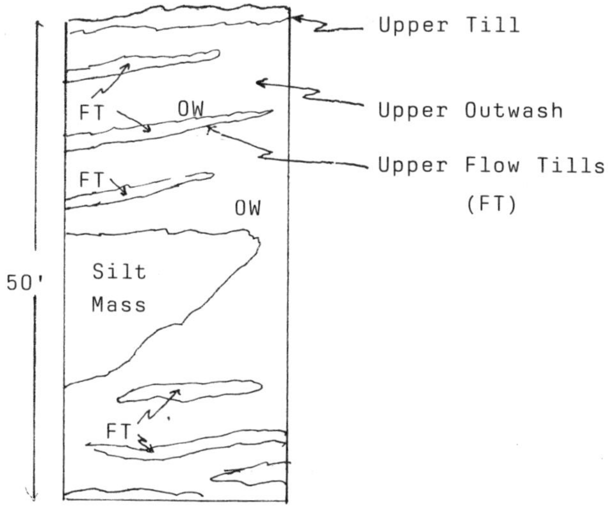


b)

Figure 39. a) Geologic Section and b) Road Cut at Roanoke Landing

The deep depression in the outwash plain north of the highway may be a kettle that has been breached by stream erosion on the south side. Numerous small kettles indicate periglacial ice melting under the outwash.

67.7 miles. Fresh Pond Landing (Wading River Quadrangle). This embayment originated as a deep, V-shaped, north-trending stream valley that was cut by north-flowing drainage. At the north end of the valley, the stream has been dammed by dominantly east to west sand bars at Fresh Pond Landing forming a marsh. The stream cuts only a thin channel through the sand bar.

69.5 miles. Turn right (north) on Hulse Landing Road. The moraine widens north to south to about 0.8 miles at Wildwood State Park and to nearly 1.2 miles through Wildwood Village, with characteristic kame and kettle topography. A traverse through Wildwood State Park will confirm this observation.

70.6 miles. Turn left (west) on North Wading River Road.

72.9 miles. North Wading River Road enters the headwaters of the Wading River valley drainage.

73.3 miles. Junction with North Country Road; continue on North Wading River Road. The moraine is nearly 1.5 miles wide north to south at Wading River.

73.5 miles. Turn right (north) on Sound Road.

74.0 miles. View of Wading River marsh and the Shoreham Power Plant on the left (west) side of the valley; the moraine forms the slope on the right (east) side.

Stop 9. 74.4 miles. Parking lot at Wading River Landing. The Wading River marsh formed in much the same way as the marsh at Fresh Pond Landing. A dominant east to west spit has blocked the drainage forming a marsh. Wading River becomes a tidal channel meandering across the marsh before entering the Sound on the west side of the spit. Groins along the beach block the drift of sand. The beach is mainly sand with quartz pebble gravel.

**FIELD TRIP 4.**

The South Shore Barrier Beaches of Eastern Long Island- Shinnecock Inlet to Moriches Inlet (Maps from east to west are: Shinnecock Inlet Quadrangle, Quogue Quadrangle, Eastport Quadrangle, Moriches Quadrangle). A working model for the present condition of the barrier beaches is the Tiana Beach-Westhampton Beach barrier between Shinnecock Inlet on the east and Moriches Inlet on the west (Figure 40), a fifteen and one half mile long barrier beach segment incorporating features of the natural barrier beach (Figure 16) and the multitude of modifications imposed on the barrier by suburban encroachment. Natural features include berms, dunes, and old washovers and inlets from ancient storms; suburban development has replaced dunes with buildings and parking lots, filled in and developed marshland, paved the interdune swale and called it "Dune Road," and emplaced fifteen sand-trapping groins on Westhampton Beach from Jessup Lane to south of Gunning Point.

The trip begins on the west side of Shinnecock Inlet at the jetty (Shinnecock Inlet Quadrangle). From Hampton Bays take Ponquogue Avenue south and follow signs to Ponquogue Bridge. Cross the bridge and take Dune Road east to Shinnecock Inlet. Park in the lot on the ocean side. Walk to the jetty. Note that after Stop 1, parking along Dune Road will be restricted until new beach parking has been completed in West Hampton Dunes.

<u>Stop 1. 0.0 miles.</u> Several coastal storms in the early 1990's took a severe toll on the barrier beach. The jetties were badly damaged and the dunes were swept into Shinnecock Bay. The

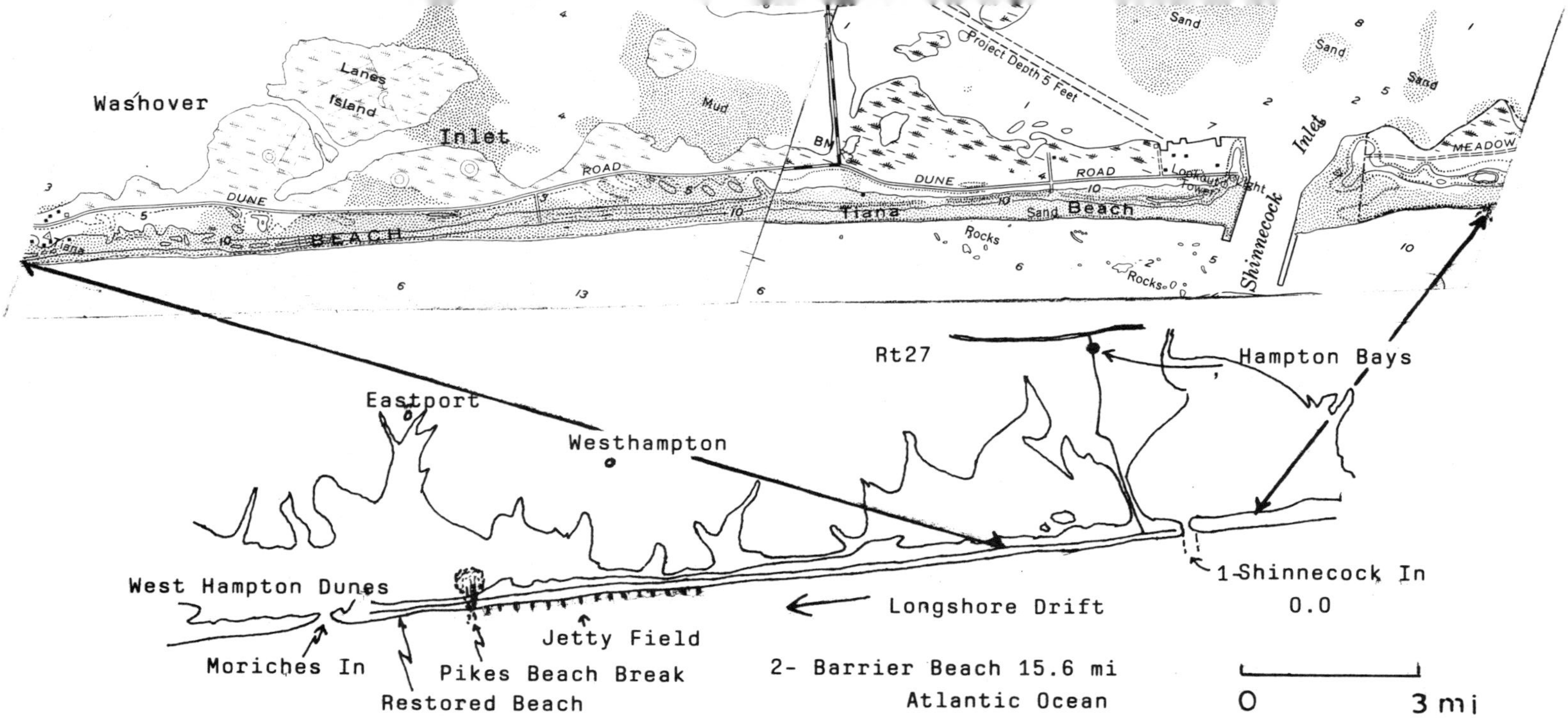


Figure 40. Field Trip 4: The Tiana-Westhampton Barrier

cumulative damage and the impact on recreational and economic interests prompted settlement of liability claims by the governmental regulatory agencies, and mitigation of the problems by the U.S. Army Corps of Engineers, New York State and local goverment agencies, and citizens organizations. The Corps is in the process of repairing the jetties with heavy blocks of rock and rebuilding the eroded beaches and roads.

The primary dunes are being replaced with a ridge of coarse sand between Dune Road and the beach front. The new sand, much coarser than the original wind-blown dune sand, is mined from the ocean floor south of the beach. Ocean waves are bent by the jetties into and around the inlet and generate seemingly- stationary waves in the narrow passage, requiring careful navigation by boaters. The jetties block the east to west longshore drift, widening the barrier on the east side of the east jetty. The jetty on the west side also bends the waves onto the beach downstream where erosion is carving a deep salient into the newly restored beach. Sand eroded from this beach is entrained in the littoral drift until it reaches the next jetty.

1.0 miles. Ponquogue Bridge (Quogue Quadrangle). North of Dune Road, fine sediment has accumulated around bridge abutments on the east side of the bridge and widened the salt marsh. These deposits reflect the east to west bay currents in this area.

2.0 to 3.0 miles. Lanes Island, north of the road, is a salt marsh built on an old washover from a prior storm. The channels east of the Island may be an old storm-cut inlet next to the washover. Such inlets are normally closed by wave-driven sand bars.

4.0 to 5.0 miles. Sedges Island is also a marsh-covered washover and is bordered by inlets.

5.0-11.5 miles. Filling of low-lying and marsh land for buildings is more evident along Dune Road in Westhampton Beach (Eastport Quadrangle). The Dune Road strip resembles a well-established and upscale residential community with resort amenities, rather than a barrier beach. Large homes are set back from the road on landscaped lots. Beach clubs and condominiums are more prevalent to the west. The characteristic features of the natural barrier are all but obliterated by development.

9.4-11.4 miles. Fifteen groins were built along this stretch of beach in the 1970's. The groins have stabilized and protected this segment of the barrier from coastal erosion, all the while trapping sand for additional beach growth.

11.4-14.0 miles. West of the groin field, the beach at West Hampton Dunes was seriously eroded by coastal storms in the early 1990's, culminating in a major nor'easter late in 1992 (Figure 41). More than 200 homes were damaged or destroyed, and most of these were washed away along with the road and other services, as well as the beach and dunes. Dune and beach sands were carried across the barrier into the bay, and new inlets were cut in the vicinity of Pikes Beach. In the aftermath of the storms, many people believed that rebuilding the area would be futile. Any future storm could be expected to repeat the destruction and remove new sand. The engineers, however, believed that by using coarser sand to rebuild the beach, including replacing the fragile dune line with a wave-resistant, ridge-like buffer zone between the forebeach and the road, the beach overall would be able to withstand storms of a greater magnitude (measured in terms of the number of years between storms of a given intensity- fifty year storms are much more powerful than five year storms, but statistically come along only every fifty years) than those experienced in this time span. They did manage to close the new inlets with new sand supplies in order to preserve the established and protected inlets at Shinnecock and Moriches. In the wake of legal action taken by the village

a) Shinnecock Inlet with Jetty Construction

b) West Hampton Dunes- Dune Road Repair

Figure 41. Views of the Barrier Beach with Repairs in Progress of West Hampton Dunes against the Federal Government, a settlement was reached in which the Corps of Engineers would rebuild the devastated areas, add to the public beach and parking facilities, and modify the groin field to allow longshore drift and sand replenishment to occur naturally. Thus, sand pumped from offshore deposits has widened the beach to over four hundred feet from the high tide line to Dune Road. A sand ridge parallel to and one hundred feet south of Dune Road forms a new backbone for the barrier beach as well as a buffer zone in which homes can be built. Additional open space will provide more protection for endangered species like the piping plover, a shorebird that nests on the open beach above the high tide line.

14.0 miles. Swan Island is an old washover, and Clear Creek is a former inlet, as is the embayment west of the Island.

14.1-15.6 miles. Cupsogue Beach County Park to Moriches Inlet (Moriches Quadrangle). Erosion of this beach segment has drawn less attention because the Park is essentially uninhabited. However, Moriches Inlet and its jetties were damaged. Washovers carried sand and salt water into the Bays, altering the bay bottom morphology and the bay salinities. This beach will benefit from the sand replacement project, and the jetty will block the passage of additional sand entrained in the littoral drift from the east and thereby feed this beach.

Stop 2. While no formal stops along Dune Road can be encouraged due to parking restrictions, brief picture stops can be made. A second stop or additional stops for this field trip should be possible where parking and beach access is available. When new beach and parking facilites are available, it will be possible to view the beach front, the inlet and jetties, the new dune ridge and other construction along the barrier by hiking along the beach.

**Field Trip 5.**

A south to north transect through the sequence of late Wisconsinan glacial deposits in central Long Island. The Connetquot River, which drains central Long Island from north to south, and the Nissequogue River, which flows from south to north, have nearly coalescing drainage basins in central Long Island (Figure 42; Maps from south to north are: Bay Shore East, Central Islip and Saint James Quadrangles). These two underfit rivers together appear to cinch the waist of Long Island and present an intriguing stream valley system that originated as a late-glacial meltwater channel draining proglacial lakes north of the terminal moraine, through the moraine, and across the outwash plain and continental shelf.

This trip straddles the arbitrary geographic boundary between eastern and western Long Island. Geologically, it is situated east of the late Wisconsinan Interlobate Zone between the Hudson and Connecticut lobes in the vicinity of Huntington. The Interlobate Zone resulted in the complex of interlobate morainal deposits, meltwater channel gravels and proglacial lake beds seen today in and around the north-south range of hills: Manetto Hills, Half Hollow Hills and Dix Hills. The route of Field Trip 5 lines up roughly with the axis of the last interlobate angle that formed between the receding glacial lobes. It trends northward toward the interlobate angle formed by the Sands Point Moraine of the Hudson Lobe and the Roanoke Point Moraine of the Connecticut Lobe. The deposits in that junction, which can be projected beneath Smithtown Bay, were eroded by meltwater flowing southward from the

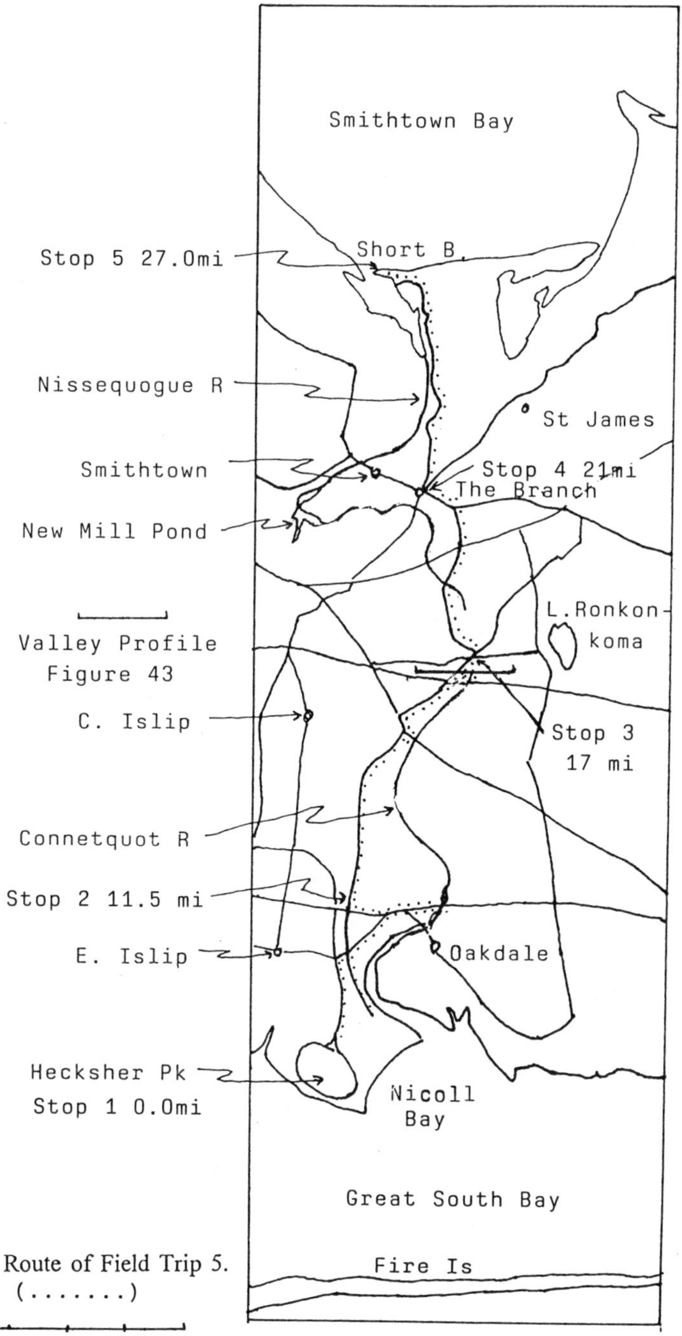


Figure 42. Route of Field Trip 5.
(.......)

proglacial lake and then submerged beneath the rising sea.

While the Connetquot and Nissequogue rivers seem to originate miles apart, as depicted on the topographic maps, their valleys begin on either side of a divide located near a broad gap in the Ronkonkoma Moraine, the terminal moraine, both at elevations of about sixty feet. They are separated by an area of pitted outwash and meltwater channel deposits formed over residual ice that is only one-fifth of a mile wide. The two valleys have meandering stream patterns and are incised north and south of the gap into outwash and lacustrine deposits. If interpreted collectively as a single meltwater channel that traversed the Island from north to south, the combined valley has a broad meander pattern that is eroded into the glacial deposits that now form its outside banks. The origin of this drainage is interpreted from the geomorphology and the sedimentary deposits encountered along the route of this trip.

Stop 1. 0.0 miles. A good starting point for this trip is at the entrance to Heckscher State Park (Bay Shore East Quadrangle). Follow the roadway loop through the park for four and one-quarter miles.

0.0-5.6 miles. The outwash plain underlies the park, the salt marshes and the bay to the south. A seemingly artificial crescent beach forms the south shore of the park. The Connetquot River empties into Nicoll Bay over one mile to the northeast, but prior to Holocene submergence the river probably meandered back to the southwest against the Heckscher shore. A few miles to the south, Fire Island forms the barrier beach that encloses Great South Bay.

5.6-5.7 miles. Turn right on Rt 27 and then right again on Great River Road. Great River Road follows the west bank on the outside curve of the last meander of the river, beginning south of Sunrise Highway. The west bank is the steep side (outside) of the meander loop, and it rises about fifteen feet above the river. The topography of the river valley is somewhat obscured by development. Turn around near golf course and proceed northward to Montauk Highway, Rt 27A.

7.9 miles. Turn right (east) on Rt 27A, essentially following the meander curve but well above the cut bank. Note that the field trip follows the meanders of the meltwater channel valley in the reverse order to that in which they formed.

9.4 miles. Junction with Sunrise Highway, Rt 27. Turn right (east) on Rt 27 and cross the Connetquot River valley near Lower Pond. The river has been dammed here to form Lower Pond and East Pond. The river bottom is partially hidden by development and highway ramps to the south. To the north, a forest of small trees masks the ponds and the channel. At the next stop light turn left- to the west- on Rt 27 and proceed westward toward Connetquot Avenue.

Stop 2. 11.5 miles. Turn right (north) on Connetquot Avenue. Stop where convenient to get oriented to the terrain. Note that while the river meanders are easily followed on the topographic map, extensive housing development hides the river valley. The road follows the western margin of the Connetquot valley, passing the Bayard Cutting Arboretum and the Connetquot State Preserve (Central Islip Quadrangle). The Connetquot River, while gently meandering on its flood plain, has historically been dammed for mill ponds, a sportsmans club and a fish hatchery. A well defined, west-trending valley meander forms the river's reach and a low eroded bank in the outwash between the Long Island Railroad tracks and the village of North Great River. But it is against the easterly bend of the valley, located in the State Preserve to the south, that meltwater once carved a twenty foot escarpment.

The breadth of the valley meanders compared with the width of the Connetquot channel shows the river to be decidedly underfit in this segment. Evidently, a much more robust, meltwater laden river established the meander pattern and shaped this much broader valley. The fifty-foot contour line and the adjacent contour lines outline the valley morphology from meander to meander northward from the State Preserve to the Ronkonkoma Moraine.

14.5 miles. Connetquot Avenue turns right.

14.5 miles. Junction with Veterans Memorial Highway, Rt 454. Bear right on Rt 454 about 0.2 miles to sign: "Headwaters of Connetquot River." The valley can be entered from a gate east of the sign with permission from the State Park Preserve, as noted. Turn back (northwest) on Rt 454.

15.2 miles. Turn right (northeast) on Nichols Road which crosses the outwash plain between the Ronkonkoma Moraine to the north and the rapidly narrowing river valley to the east.

15.6 miles. Note the water recharge basin to the left.

16.5 miles. Cross the Long Island Expressway.

Stop 3. 17.0 miles. Roughly where Nichols Road, Terry Road and Vanderbilt Motor Parkway converge in a triangular intersection, about 0.4 miles north of the Long Island Expressway, Rt 495, the headwaters of the Connetquot River are now buried in drains under the roadways. The basin begins in a significant gap in the Ronkonkoma Moraine. Stop where convenient near the gap. The late-glacial meltwater channel was incised through the moraine as meltwater carved an easterly meander sixty feet above present sea level. In a short excursion to the east, Vanderbilt Motor Parkway ascends the Ronkonkoma Moraine. To the west, the Parkway crosses the gap and climbs the meander-scarred moraine (Figure 43).

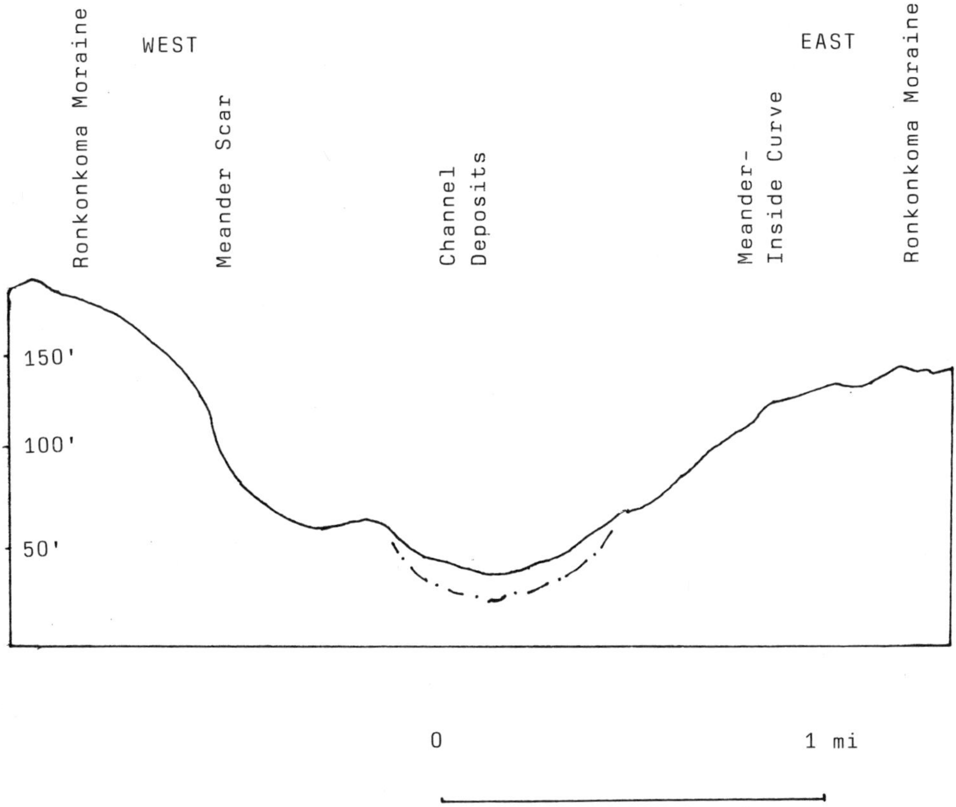


Figure 43. Profile of Meltwater Channel Meander at Ronkonkoma Moraine Gap

17.1 miles. Turn left (northwest) on Terry Road. The road goes up a hill that follows the meander scar. The evidence of meander erosion shows that meltwater flowed at this higher elevation during the interval of valley widening. Upstream, the valley turns sharply from the northwest against the steep proximal slope of the moraine, with a seventy foot-high meander carved into it.

The northern half of this meltwater-formed meander has become, in the context of the modern drainage, the headwaters of the north-flowing Nissequogue River drainage, a region of wetlands and ponds within the fifty-foot contour line. The river valley casts two broad meanders over the relatively flat basin enclosed by the Ronkonkoma Moraine to the south and the recessional moraines of the last glaciation to the north (the Stony Brook Recessional Moraine of the Connecticut Lobe of the glacier to the northeast and the Northport Recessional Moraine of the Hudson Lobe to the northwest; see Figure 5). This lowland, characterized by meanders incised into a plain of low relief, wetlands and pitted outwash, is underlain by lake clays deposited on the floor of a proglacial lake formed between the terminal moraine and the ice front during deglaciation.

Continue northward on Terry Road from the intersection.

18.1 miles. Intersection with Townline Road. Terry Road enters the hummocky topography of the moraine.

19.1 miles. Intersection with and merge left into Smithtown Blvd.

19.6 miles. Intersection with Smithtown By-Pass; continue on Terry Road.

20.0 miles. Note the exposure of a kame along the roadside.

20.2 miles. Another kame can be seen to the north. Note erosion of sand and gravel forming slides on the slope.

20.4 miles. Intersection with Middle Country Road, Rt 25. Turn left (west) on Rt 25.

Stop 4. 21.0 miles. Junction of Rt 25 with Rt 111, Rt 25A and Nissequogue Road in the village of The Branch, just east of Smithtown. Stop where convenient to get your bearings and consider these features. Terry Road and this segment of Rt 25 follow the edge of a proglacial lake basin that formed between the Ronkonkoma Moraine and the receding ice front during the late Wisconsinan. The roads curve against the meander-cut valley wall. The underfit channel of the Northeast Branch of the Nissequogue River first wanders northward through (former) wetlands before making a broad northward curve.

The channel once included several small ponds formed from dammed-up stream segments. It then heads westward, south of Smithtown, to New Mill Pond, also the result of a dam, and then turns northward across the flat glacial channel bottom and flood plain deposits where the valley narrows significantly. While sufficiently clear on the topographic map, the channel is obscured on the east side by housing, office buildings and a shopping center. The poor drainage on the proglacial lake-bottom clay beds, referred to as the Smithtown Clay, of late Wisconsinan age, serves as a reminder of the geologic history of this lowland. Emerging from the lake basin to the southwest of Smithtown, the valley turns northeastward; the narrow river channel incised into the outwash with high banks on either side. The river can be seen south of and parallel to Rt 25, west of the Rt 25-Rt 25A split west of Smithtown, where it flows northeastward through a young hardwood forest established on the flood plain. It then crosses Rt 25 just east of the Rt 25-Rt25A split, about one mile west of the Rt 111 intersection.

The narrow valley persists through a broad eastward meander north of Smithtown where the meltwater channel cut through the recessional moraine. The modern Nissequogue River actually meanders in a confined belt on its own marsh-covered flood plain, but now with the steep valley walls of the meltwater channel up to one hundred feet high. The river becomes tidal near Mill Creek, and tidal flats cover channel deposits northward to the river mouth. Where the Nissequogue River enters Smithtown Bay, it is restricted by sand bars from both the east and west sides. The east side's northwest-hooking spit, Short Beach, reflects the dominant direction of littoral drift. It has forced the river outlet to bend further westward; the west sand spit has, in turn, confined the westward discharge.

21.1 miles. Immediately after crossing the intersection take the left fork, Nissequogue River Road, and proceed northward (Saint James Quadrangle). Pass the old cemetery on the right.

21.6 miles. Within a short distance, the road rises onto the distal slope of the recessional moraine, the Stony Brook Moraine.. The river lies near the interlobate angle, but Nissequogue River Road and adjoining roadways cut through successive ridges of the Stony Brook Moraine for over three miles between here and the north shore. This segment of the Stony Brook Moraine is typically hummocky and has numerous kettles.

23.3-23.6 miles. A good view of the river to the west (left) and the eroding roadcut into the moraine on the east (right). Merge with River Road.

24.0 miles. Views of Nissequogue tidal estuary to west.

25.6 miles. River Road intersects Moriches Road, then becomes Horse Race lane, and then Boney Lane at 26.1 miles.

26.8 miles. The road descends the kame moraine to the level of the salt marsh, and then enters Smithtown Short Beach.

Stop 5. 27.0 miles. Park and walk to the beach. Consider the variety of coastal landforms (Figure 44). To the southwest is a bluff cut into the moraine by coastal processes. To the northeast you can see the topography of the Roanoke Point Recessional Moraine forming the distant bluffs on the Old Field headland. The successive ridges of this eroded westernmost extension of the moraine are apparent from this perspective. Try to imagine the confluence of the Roanoke Point and Sands Point moraines, to the north in what is now Smithtown Bay, as it might have looked in late-glacial times before it was erased by meltwater streams and coastal currents. To the east, large erratics in the Bay suggest the most recent increment of sea level rise and erosion of the moraine. Today bay currents are dominantly east to west along this beach.

From the parking lot, follow the dune ridge westward to the River and the boat ramp.

27.2 miles. Here you can observe the estuary and the meanders of the meltwater channel valley south of the Short Beach spit (Figure 44)

a) Bouldery Fishers Island Moraine

b) Erratics Offshore- Rocky Point

Figure 44. Views of the Beach, River and Moraine

c) Kame Delta- Fishers Island

d) Nissequogue River Estuary

Figure 44. Views of the Beach, River and Moraine Continued

e) Roanoke Point Moraine Bluffs

f) Roanoke Point Moraine and Gravel Berm

Figure 44. Views of the Beach, River and Moraine Continued

## GLOSSARY

**Alluvium.** Sediment deposited by streams.

**Amphibolite.** A rock type, usually metamorphic, mainly of iron, magnesium silicate minerals, such as hornblende and plagioclase feldspar.

**Appalachian Mountains.** The prominant mountain ranges of varying ages that make up much of eastern North America.

**Aquiclude.** A confining bed for an aquifer.

**Aquifer.** A porous and permeable layer of rock or sediment capable of transmitting groundwater.

**Archean Era.** The most ancient major division of geologic time, greater than 2.5 billion years.

**Avalanche.** See soil flow.

**Avalonia.** A large crustal fragment, mainly an island arc-basin complex that merged with Ancestral North America early in the Paleozoic.

**Back Beach.** The area of the beach landward of the berm.

**Barchan Dune.** A crescent-shaped sand dune whose outer edges, or horns, advance to leeward and embrace the short, steep leeward slope within the concave arc. The longer slope is convex to windward.

**Barrier Beach.** An elongate, offshore sandbar complex deposited parallel to the shoreline in a down-current direction. Barriers isolate irregular features along the coast to form bays, estuaries, lagoons, and ponds.

**Barrier Beach Aquifer.** The usually unconfined groundwater lens within a barrier beach.

**Basalt.** A fine-grained igneous rock, basic in chemistry and composed of plagioclase feldspar and iron and magnesium silicate minerals. Rocks of this general description are referred to as basaltic rocks.

**Basal Till.** A compact, dense till formed in the basal zone of glaciers. See also Lodgement Till.

**Batholith.** A mountain-sized, igneous intrusion often formed in continental crust above a subduction zone where high heat flow generates magma.

**Beach.** A seaward sloping, convex upward, shoreside plain that is usually covered with sediment.

**Beach Cusp.** See cuspate berm.

**Beach Face.** The seaward sloping portion of the beach plain in the tidal zone.

**Bedrock.** The solid rock of the earth's crust, often exposed beneath unconsolidated sediments.

**Bench.** A relatively level erosion plane above the surrounding land surface, often forming a low platform above sea level where associated with coastal erosion.

**Berm.** On a beach, a raised depositional ridge at the high tide level of the beach face, marking the division between the forebeach and the back beach areas.

**Blowout.** An erosion hollow formed on open ground or among sand dunes by prevailing winds.

**Blowout Dune.** Sediment from the deflated area may form this type of sand dune on the leeward side of the blowout. The dune may resemble an inverted barchan with horns wrapping around the depression.

**Bluff.** A high bank or eroded headland with a steep slope or cliff rising above a river plain or along a coast.

**Bog.** A wetland, often with a thick accumulation of peat below an active marsh.

**Bottomset Beds.** See Delta.

**Cambrian.** The initial geologic period and system of rocks of the Paleozoic.

**Causeway.** A roadway built upon fill above an unsuitable substrate, such as a beach or a marsh.

**Cenozoic Era.** The most recent of the major divisions of geologic time and divided into the Tertiary and Quaternary periods.

**Climatic Refuges.** See Refugia.

**Coastal Plain.** A relatively flat or gently sloping plain of layered sedimentary strata that generally merges with the continental shelf.

**Cobbles.** Small, rounded rock fragments from three to twelve inches- larger than gravel and smaller than boulders.

**Collapse Structures.** See ice-contact structures.

**Concretion.** A nodular mass of locally-formed compacted or encrusted sediment, generally incorporated in sedimentary strata of similar age but contrasting grain size.

**Confined Aquifer.** An aquifer in which the water transmitting layer lies below or between relatively impermeable layers.

**Conglomerate.** Sedimentary rock composed of coarse-grained, rounded rock fragments in a fine-grained matrix.

**Connecticut Lobe.** One of the main salients, or lobes, of the late Wisconsinan glacial margin in southern New England, controlled by the bedrock and morphology of the Connecticut River

Valley.

**Connecticut-Rhode Island Sublobe.** A minor lobe of the late Wisconsinan glacial margin in southern New England between the Connecticut Lobe and the Narragansett Lobe.

**Continent.** A major landmass, often comprising a single crustal plate along with the adjoining oceanic crust.

**Continental Shelf.** The gently-sloping subsea margin of the continent.

**Convection.** The mode of heat transfer and material movement in low viscosity materials. Convective flow in the earth's upper mantle is the driving mechanism of plate tectonics.

**Coral.** A marine invertebrate, usually colonial, whose external skeletons are attached in large numbers to form reefs, usually in sunlit, tropical seas.

**Corps of Engineers.** U.S. Army agency charged with regulating and maintaining river and coastal zone structures, as well as military installations.

**Country Rock.** Existing bedrock into which igneous intrusions may be emplaced and which is subject to contact and regional metamorphism.

**Crescent Beach.** An arc-shaped beach formed as a sand bar or infilling of an embayment that connects two headlands.

**Cretaceous.** The third and last geologic period of the Mesozoic Era; rock units formed during that time.

**Cross-Bedded Sediments.** Inclined laminations of sediment within and at an angle to the main, incorporating bed of sediment.

**Crust.** The outer layer of the earth above the mantle; it is segmented into plates which carry continents and ocean basins.

**Crustal Plate.** A division of the earth's crust; one of a number of adjacent, major and minor continental and oceanic crustal segments.

**Cuspate Berm.** A series of crescentic troughs carved into a berm by the action of waves. Also known as beach cusps.

**Debris Flow.** See soil flow.

**Delta.** A landform resulting from sediment deposited by a stream or river into a larger body of water, such as a lake or ocean. Deltas are typically triangular in plan and are comprised of three sets of sedimentary beds: the topsets represent deposits in the stream's flood plain; the foresets form on the depositional slope below water level; and the bottomsets represent the fine sediments that settle onto the floor of the lake or sea.

**Deposition.** The process of laying down sedimentary particles by running water, glaciers, and wave, wind or coastal currents.

**Distal Slope.** The slope or side of a moraine facing away from the glacier.

**Divide.** The upland or interfluve separating drainage basins.

**Drift Sheet.** Glacial deposits in a region associated with one glaciation. On Long Island, the drift sheets can be designated the lower, or older drift, and the upper, or younger, drift.

**Drumlin, Drumlin-shaped Hill.** A glacially streamlined, oval-shaped hill, often resulting from glacial advance over older glacial deposits, other sediments or erodable bedrock.

**Dune.** A hill of wind-blown sand; a sand deposit bearing distinctive, gently-sloping windward layers and steeply-sloping leeward beds.

**Dune Line.** Lineations of dunes in rows parallel to the source area. See Primary Dune, Secondary Dune.

**Dune Ripple.** Miniature dune forms in parallel ridges covering the surface of a dune.

**End Moraine.** The moraine deposited at the farthest advance of a given glacier or glacial readvance.

**Entisol.** A soil characterized by a lack of distinct horizons.

**Epoch.** A subdivision of a geologic period.

**Era.** A major subdivision of geologic time.

**Erosion.** The process of removing and transporting materials through the action of wind, running water, glaciers, waves and coastal currents.

**Erratics.** Cobbles and boulders transported, usually by a glacier, from their source areas into regions of dissimilar bedrock.

**Escarpment.** A more or less continuous cliff formed by erosion or faulting. Also known as a Scarp.

**Estuary.** The lower reach of a stream that has become drowned and subject to coastal currents, such as tides, and increased salinity due to subsidence of the land or sea level rise.

**Eurasia.** An early Paleozoic supercontinent incorporating Europe and Asia.

**Fault.** A fracture in bedrock or layers of sediment along which there is relative displacement of the two sides.

**Feldspar.** A group of common, rock-forming aluminum silicate minerals.

**Ferromagnesian.** Term generally applied to rocks rich in iron and magnesium silicate minerals, such as amphiboles and pyroxenes.

**Fetch.** The direction from which waves are propogated by winds blowing over open water.

**Flood Plain.** The accumulation of sedimentary layers that make up the deposits of a river in its valley. The river continuously reworks and modifies the surface of the flood plain. See Meanders.

**Flow Till.** Till washed as a mass of sediment from melting glacial ice, and often interlayered with outwash.

**Foliation.** Lamination resulting from the segregation of light and dark minerals in metamorphic rocks.

**Forebeach.** The area of the beach seaward of the berm.

**Foreset Beds.** See Delta.

**Gardiners Clay.** Stratigraphic term applied to fossiliferous, marine sands and clays found thrusted into the recessional moraine on Gardiners Island, and to a sequence of clay deposits found below sea level south of the Ronkonkoma Moraine in southeastern Long Island. Both units are thought to represent deposits that originated during the last interglacial. However, their sediment and fossil contents vary considerably, and there are no confirming radioimetric ages to support this idea.

**Glacial.** Features and conditions characteristic of glaciers.

**Glacial Advance.** Movement of a glacier, radiating outward from a center of ice accumulation, down a mountain valley into lower elevations or over a continental region into lower latitudes.

**Glacial Aquifer.** The water-transmitting and generally unconfined layers of a glacial drift sheet, such as the "upper glacial aquifer" of Long Island.

**Glacial Deformation.** Reorganization of masses of sediment, including older units as well as contemporaneous glacial deposits by glacial movement. Material may be shoved, faulted, folded, rafted and redeposited, or thoroughly blended by the advancing ice. It can also be slumped or collapsed due to the melting of supporting ice.

**Glacial Drift.** Sediment carried and deposited by a glacier.

**Glacial Lake.** A body of water resulting directly from glacial melting. It may be impounded between a glacier and adjacent bedrock or moraine, or it may occupy glaciated terrain. See Proglacial Lake.

**Glacial Lobe.** A regional extension or salient of a portion of an ice front whose direction of movement may be controlled by the underlying topography and geologic structure.

**Glacial Readvance.** Renewed forward movement of the ice, usually due to climatic cooling and the resulting increase in the accumulation of snow and ice on the glacier; otherwise due to internal dynamics of the ice mass. Often invoves ice shove and deformation, reorganization and drumlinization of existing deposits.

**Glacial Recession.** Melting of an ice sheet, usually due to climatic warming, when melting of the ice exceeds resupply of snow; generally causes downward wasting and thinning of the ice mass, and retreat of the ice margin.

**Glacial Stage.** A major interval of glaciation within a glacial epoch.

**Glacial Stand.** Temorary state of equilibrium of the ice front at which time the margin is stationary. Also known as a Stillstand.

**Glacier.** An accumulation of compacted snow and ice in a mountain valley or in a polar land area that develops the ability to flow down grade or laterally due to its own weight, recrystallization of the ice, and other internal dynamics.

**Gneiss.** A foliated metamorphic rock in which visible mineral grains are arranged in light and dark bands.

**Gondwana.** The supercontinent of the southern hemisphere during the early Paleozoic and post-rifting Mesozoic eras; the supercontinent is comprised of continental plates ancestral to Africa, South America, India, Australia and Antarctica.

**Granite.** A medium-grained, igneous rock of acidic chemistry, made of quartz, feldspar, mica, and accessory minerals.

**Granite Gneiss.** A gneiss in which the foliations or bands contain granitic minerals; it often closely resembles granite.

**Granitic.** Pertaining to granite and related rock types, the mineral composition of granite, and the typical composition of the continental crust.

**Greenhouse Effect.** The ability of the atmosphere to trap heat energy reradiated by the earth, much like the glass of a greenhouse.

**Groin.** A jetty-like structure constructed perpendicular to a beach in order to trap sediment in the littoral drift, nourish a beach, and prevent erosion.

**Ground Moraine.** A term usually applied to nonlinear, intermorainal deposits of glacial drift.

**Groundwater.** Water that has infiltrated the soil, bedrock or an aquifer forming a water saturated zone whose upper surface is known as the groundwater table or water table.

**Groundwater Divide.** Subsurface topographic or structural high from which groundwater flows.

**Groundwater Reservoir.** An aquifer or zone of groundwater saturation in bedrock or sedimentary layers.

**Groundwater Surface.** The upper boundary of the zone of groundwater saturation; the water table.

**Gullying.** Formation of deep erosion channels by running water prior to the development of a stream valley.

**Hartford Basin.** A Mesozoic rift valley filled with red beds and basaltic lavas; now occupied by the Connecticut River valley.

**Herb Pollen Zone.** See Pollen Zones.

**Headland.** A promontory of land usually made of resistant rock or sedimentary layers, located along a river or coast and projecting into the water.

**Holocene.** The most recent and current geologic epoch and the youngest epoch of the Quaternary Period, dating from 0 to 10,000 years ago; also known as the Recent Epoch.

**Hoodoos.** Pillars and spires that are the erosional remnants of compact sedimentary layers; often found in sandstone and lodgement till.

**Hook.** The end of a spit that curves due to deflected coastal currents.

**Horizon.** A distinctive, thin bed of rock or sediment, or a unique surface marking a boundary within the bedrock or sediment layers.

**Hudson Lobe.** The major lobe of the late Wisconsinan glacial margin in eastern New York controlled by the bedrock and morphology of the Hudson River valley.

**Hummocky Topography.** The irregular, depositional surface of glacial moraines, with hills (knobs or kames) and depressions (kettles); also known as kame and kettle topography.

**Ice Blocks.** Masses of ice calved from a glacier usually into surrounding sediments, such as outwash, adjacent to the ice.

**Ice-Contact Structures.** Glacial deposits adjacent to the ice may be faulted, folded or otherwise deformed by glacial movement or melting of buried ice.

**Ice-cored Glacial Deposits.** Glacial or fluvial sedimentary layers deposited over buried ice.

**Ice Front.** The down-ice extremity of a glacier.

**Ice Margin.** The temporarily stable, regional extremity of a glacial front that results from the balance between resupply of snow to the glacier and its melting and flow.

**Igneous Intrusion.** Movement of molten rock (magma) from a magma chamber into the country rock.

**Igneous Rock.** A type of rock formed from the cooling and crystallization of magma and lava.

**Illinoian Glacial Stage.** The third of four major glacial stages of the Pleistocene Epoch.

**Impermeable Sediments.** Layers of sediment of relatively low ability to transmit fluids.

**Indian Paint Pot.** A rock-like vessel or concretion formed through deposition of iron oxides around a ball of clay. When hollowed out, these have been used as paint pots.

**Interfluve.** The upland between stream valleys.

**Interglacial.** The warm climatic stage between glacial stages.

**Interlobate.** Pertaining to the area where two glacial lobes merge.

**Interlobate Angle.** The angle formed by the margins of two glacial lobes; the angle formed by intersecting moraines deposited by two intersecting lobes.

**Interlobate Moraine.** Moraine deposited in the area between or in the intersection of two glacial lobes.

**Interlobate Zone.** The complex of deposits and landforms formed between two glacial lobes.

**Intermittent Drainage.** Drainage in which streams flow only part of the year.

**Interstadial.** A warm climatic interval between ice advances within a glacial stage.

**Inverted Topography.** Topography in which a geological feature normally of depressed elevation, such as a lake bed, is elevated above the surface as the result of erosion.

**Iron-Cemented Zone.** Groundwater channels formed in sedimenary layers often have iron compounds deposited in the form of pipes, lenses and zones.

**Iron Pyrite.** The gold-colored mineral, pryite (iron sulfide, FeS2); also known as fools gold.

**Island Arc.** A chains of islands, usually volcanic in origin.

**Island Arc-Basin Complex.** A volcanic island arc and the depositional basins adjacent to it, similar to the Islands of Japan and the Sea of Japan.

**Isotope.** Any of two or more forms of the same element, having the same or closely related properties and the same atomic numbers but different atomic weights.

**Isthmus.** A narrow body of land connecting two larger land masses.

**Jetty.** An elongate structure, usually made of large, quarried blocks of rock, that is constructed perpendicular to the shore and sometimes in pairs and adjacent to an inlet, river mouth or harbor in order to protect the passage from storms or siltation.

**Jurassic.** The second and middle period of the Mesozoic Era; the rock units formed at that time; the "Age of Dinosaurs."

**Kame.** An ice-contact deposit in the form of a conical hill of stratified outwash formed by meltwater flowing off the ice or into a crevasse in the ice.

**Kame and Kettle Topography.** Hummocky morainal topography containing kames (hills) and kettles (depressions).

**Kame Delta.** Similar to a kame, but deposited as a delta into standing water adjacent to the ice.

**Kettle.** A depression formed in glacial drift by the melting of a remnant block of ice.

**Kettle Lake.** A lake formed in a kettle; usually the base of the depression becomes sealed by fine-grained sediment as the ice block melts, allowing for water and sediment to collect.

**Lag Deposit.** Coarse-grained rock fragments that remain after fine-grained materials have been eroded away.

**Laurasia.** The northern hemisphere supercontinent of the early Paleozoic, incorporating continental blocks ancestral to North America, Europe and Asia.

**Laurentide Ice Sheet.** The name given to the portion of the continental glacier, or ice sheet, that formed in eastern Canada during the last glaciation; the ice sheet that advanced over New York and New England.

**Lava.** Molten rock from inside the earth that reaches the surface through faults, fissures, rifts and volcanoes.

**Lava Flow.** A layer of molten rock that flows onto the surface.

**Leachate.** A liquid that flows through some materials, for example, the polluted liquids that drain from a landfill.

**Lens.** A body or layer of material that is thicker in the middle and tapered at its edges.

**Lignite.** A brown to black variety of soft coal.

**Little Ice Age.** A world-wide cold climatic episode between the 15th and 19th centuries.

**Littoral Drift.** See Longshore Drift.

**Loam.** A soil made of varying amounts of clay, sand, silt, and organic debris.

**Lodgement Till.** Till deposited beneath glacial ice and compressed and compacted by the weight and movement of the ice.

**Loess.** Silt winnowed from recessional glacial deposits by prevailing winds from the ice. The silt is deposited downwind in structureless, homogeneous layers of varying thickness.

**Longitudinal Dune.** A dune ridge elongated parallel to the dominant wind direction.

**Longshore Current.** The current generated by the prevailing direction of waves at an angle to the shoreline.

**Longshore Drift.** The transport of sediment down current and parallel to the shore as a result of the longshore current. Also known as Littoral Drift.

**Magma.** Molten rock beneath the earth's surface.

**Magothy Aquifer.** Water transmitting strata generally incorporated in the Upper Cretaceous Magothy Formation.

**Mantle.** The layer of the earth between the crust and the core, consisting of peridotite, a rock type that is ultrabasic in chemistry and denser than basalt.

**Mass Wasting.** Downslope movement of rock debris in response to gravity and aided by water content; it varies in rate and magnitude depending on the slope, rock materials and water saturation.

**Meanders.** The characteristic pattern of stream flow over flood plain deposits in which the stream first forms a gently sinuous course; the curves then gradually broaden into more exaggerated convex loops. The meanders migrate across the flood plain and downstream as the stream erodes against the outside bank and fills in the inside of the loop with sand bars.

**Meander Pattern.** The complex of meander loops and related landforms and deposits that occupy the width of the flood plain. Also known as the Meander Belt.

**Meltout Till.** Till deposited during wasting of the glacier; often of a stony texture with some of the fine sediment washed away.

**Meltwater Channel.** A channel cut by glacial meltwater, usually into glacial sediments and draining an ice margin or proglacial lake.

**Mesozoic Basin.** A rift valley or basin that formed near the present continental margins during the rifting of Pangaea.

**Mesozoic Era.** The major division of geologic time between the Paleozoic and Cenozoic eras; this era is represented by three periods: the Triassic, Jurassic and Cretaceous.

**Metaconglomerate.** A metamorphosed conglomerate in which some recrystallization of the matrix has taken place and rock fragments have been sheared and stretched.

**Metamorphic Rock.** A type of rock resulting from the chemical and physical alteration of existing rock types by heat, pressure and chemical change. It involves recrystallization and growth of minerals, compaction and increase in density, and the creation of shear planes or foliations along which minerals grow.

**Mica.** A group of common, rock-forming silicate minerals that have a platy structure.

**Mid-Wisconsinan.** A reference time for events happening between early and late Wisconsinan glaciations; the term has been applied to a warm interval centered around 30,000 years ago.

**Montauk Till.** The name generally applied to the till deposits in the lower drift sheet of eastern Long Island.

**Morainal Topography.** Pertaining to the morphology of the moraine which is usually irregular and characterized as hummocky or kame and kettle topography.

**Moraine.** A linear ridge of glacial sediment marking the position of an ice margin.

**Mud Flow.** See Soil Flow.

**Narragansett Lobe.** A major lobe of the late Wisconsinan glacial margin in southeastern New England controlled by the bedrock and morphology of the Narragansett Embayment.

**Newark Basin.** A Mesozoic rift basin in northeastern New Jersey that is filled with red beds and basaltic lavas.

**Nor'easter.** Contraction of "northeaster." Describes a cyclonic storm whose position off the New England coast generates strong winds and waves from the northeast that usually result in coastal flooding and erosion.

**Oak Pollen Zone.** See Pollen Zones.

**Obduction.** During plate collisions, fragments of the descending plate may attach to the upper or overriding plate.

**Ocean Basin.** The ocean floor beyond the continental shelf.

**Oceanic Crust.** The earth's crust beneath the ocean; generally composed of rock of basaltic composition.

**Oceanic Plate.** A crustal plate underlain by oceanic crust.

**Outwash.** Layered sediments, usually sand and gravel, transported and deposited away from the glacial margin by meltwater.

**Outwash Fan.** The fan-shaped bodies of outwash adjacent to the distal slope of a moraine.

**Outwash Plain.** Outwash from a glacier may form a broad plain adjacent to the ice margin or end moraine.

**Paleosol.** A soil horizon, surviving from a prior soil-forming interval; it usually has been buried and preserved by younger sediments.

**Paleozoic Era.** The major division of geologic time before the Mesozoic Era and after the

Proterozoic Era; it includes geologic periods from Cambrian to Permian.

**Pangaea.** The ultimate supercontinent of the late Paleozoic and early Mesozoic eras, made up of all known land masses joined together during early and mid-Paleozoic time. Pangaea began to break up at the end of the Triassic Period. Its fragments form the present continents.

**Parabolic Dune.** A type of dune concave to the wind. See Blowout Dune.

**Pater Noster Lakes.** A chain of lakes formed during glacial recession, often as kettles and associated with meltwater channels and ice-cored outwash.

**Peat.** Organic sediment made of the partially decomposed remains of wetland plants, compacted and chemically altered and compressed into layers; generally forms the strata of bogs.

**Peat Bog.** A type of wetland characterized by thick accumulations of peat.

**Pedalfer.** Forest soils of humid climates that are enriched in aluminum and iron. See Podzol.

**Pegmatite.** An igneous rock type, usually of granitic composition, characterized by very large crystals which result from slow crystal growth in cool magmatic solutions. Minerals include quartz, feldspar, mica, and many rare minerals.

**Perched Water Table.** The grounwater surface formed in strata above an impermeable layer that lies above the regional water table.

**Peridotite.** A general name for the denser, iron and magnesium-rich rock, composed of the minerals olivine and pyroxene, believed to be characteristic of the earth's mantle.

**Periglacial.** Processes and conditions in terrain near a glacial margin.

**Period.** The major geologic time unit of an era.

**Permafrost.** A zone of permanent ice or frozen ground beneath the surface of the earth generally found in polar or glaciated regions.

**Permeable.** The characteristic of rock and sediment layers that allows the transmission fluids through interconnected pore spaces.

**Phyllite.** A low-grade metamorphic rock, similar to slate, but with a sheen due to the growth of microscopic, platy minerals like mica.

**Pine Barrens.** Sandy, well-drained, but low in organic nutrients, terrain, such as an outwash plain, that supports a low forest of pitch pine and scrub oak.

**Pine Pollen Zone.** See Pollen Zone.

**Pipe, Piping.** The tendency for groundwater to form channels in sediments. Redeposition of iron oxides on the walls of these channels forms pipe-like or piping structures.

**Pitted Outwash.** An outwash plain adjacent to the ice margin or moraine with numerous kettles formed by the melting of ice blocks or ice-cored sediments.

**Plagioclase Feldspars.** A group of common rock-forming minerals composed of sodium and/or calcium, aluminum silicates, and having very similar physical properties.

**Plate.** See Crustal Plate.

**Plate Tectonics.** The processes associated with the interaction of the earth's crustal plates and the upper mantle, including plate collisions, mountain building, subduction, rifting and development of island arc-basin complexes.

**Pleistocene Epoch.** The older of the two epochs of the Quaternary Period, 10,000 to 2 million years ago; the Holocene, 0-10,000 years ago, is the younger epoch; an interval of cyclic glacial and interglacial stages.

**Plume.** The extent of seepage of leachate into subsurface strata; also applies to upwellings of mantle rock on convection currents.

**Podzol.** Soil with an organic mat, a leached A zone and a B zone enriched in iron and aluminum oxides, and developed under forests in cool, humid climate. Related terms: Spodosol, Pedalfer.

**Pollen Diagram.** A graphical chart that depicts the percentage of tree and non-tree pollen varieties in sediment samples. When taken at specific depths from bog or lake cores, the diagram provides a record of vegetational change over time as interpreted from the changing pollen assemblages, or pollen zones, and associated radiometric dates.

**Pollen Zones.** The zonation is a means of synthesizing data derived from the abundance of fossil pollen in assemblages in sequential sediments. The zones approximate the general vegetation at a given level or time of deposition from the percentage of pollen types. In Quaternary sediments, such as lake and marine clays, peat bogs and paleosols, pollen zones are useful in reconstructing pre- and postglacial vegetation and climate. Postglacial pollen zones range from the Herb Pollen Zone representing tundra vegetation and cold climate; through the Spruce Pollen Zone, an interval during the late glacial when spruce forests succeeded tundra on recently deglaciated ground; the Pine Pollen Zone, a warmer and dryer episode when pine replaced spruce, especially on well-drained glacial sediments; and the Oak Pollen Zone of the Holocene, when oak forests succeeded regionally in response to cooler and more humid conditions.

**Pollution.** Addition of toxic substances from domestic, commercial and industrial wastes to the atmosphere and the earth's surface and groundwater resources.

**Pore Space, Porosity.** The open spaces between mineral grains in sediment or rock; connected pore spaces allow fluids to move through the porous material. See Permeability.

**Portwashingtonian Warm Interval.** A mid-Wisconsinan Stage, warm episode named after the geographic location of the characteristic fossiliferous and radiocarbon-dated sedimentary deposits in northwestern Long Island, New York.

**Precambrian.** The four billion years of geologic time before the Cambrian Period (the earliest period of the Paleozoic, beginning 590 million years ago).

**Precambrian Shield.** The relatively stable, ancient continental crust of precambrian age that forms the core of a continent.

**Primary Dunes.** The first line of sand dunes adjacent to the back beach, usually the most-recently formed dunes.

**Proglacial Lake.** A lake that forms adjacent to a glacier, often confined by the ice front, valley walls and moraines.

**Proterozoic Era.** The major division of geologic time between the younger, Paleozoic Era, and the most ancient, Archaen Era.

**Protocontinent.** The initial continental crust to form early in the earth's history and to be differentiated chemically from oceanic crust.

**Provenance.** The source rock or terrain for transported sediments, as revealed through analysis of similar rock and mineral varieties.

**Proximal Slope.** The slope of a moraine facing the glacier (the opposite of the distal slope).

**Pyrite.** See Iron Pyrite.

**Pyroxene.** A group of common, rock-forming minerals composed of iron and magnesium silicates.

**Quartz.** A common rock-forming mineral, as a major constituent of granitic rocks, composed of silicon dioxide.

**Quartzite.** Term applied to rock types made of the mineral, quartz. These include the metamorphic variety often found in veins and fracture fillings; metamorphosed sandstone in which the quartz grains have been fused by heat and pressure; and relatively pure, quartz sandstone.

**Quaternary.** The youngest and current period of the Cenozoic Era, 0-2million years ago, and following the Tertiary Period. It is comprised of two epochs, the Pleistocene and Holocene.

**Quick Clay.** Water-saturated clays that may be mobilized by some type of shock or disturbance. See Soil Flow.

**Radiocarbon Dating.** A method of determining the absolute age-up to about 40,000 years ago- of earth materials bearing radioactive carbon (C14) by using the rate at which the isotope decays to a stable form.

**Raritan Formation.** The initial stratigraphic unit of the Upper Cretaceous and comprised of sand and clay layers that act as confining zones and aquifers.

**Recent.** See Holocene.

**Recessional Moraine.** Moraine formed during a temporary stand of the ice front in the overall process of glacial retreat.

**Recessional Topography.** Characteristic landforms created during glacial retreat through melting and runoff, deposition and erosion; including such features as kames, kame deltas, deltas, drumlins, kettles, moraines, meltwater channels and outwash plains.

**Refugia.** Local regions where flora and fauna under climatic stress may be able to survive.

Also known as Climatic Refuges.

**Relict.** A remant feature of an earlier geologic event.

**Ridge and Swale Topography.** A depositional surface characterized by parallel depositional ridges and eroded shallow depressions. Also known as Ridge and Runnel Topography.

**Rift.** A regional fault that runs parallel to the regional structure, for example, a plate boundary.

**Rift Basin.** A down-thrown or tilted crustal segment parallel to a rift in which water and sediment may collect.

**Rift Valley.** Similar or synonymous with Rift Basin, but linear and often bounded by two parallel faults.

**Rip Ups.** Pieces of underlying bedrock or sediment layers that are detached and thrusted by overriding forces, such as glacial ice. The detached and thrusted materials may be incorporated in the younger strata.

**Rotational Plane.** A plane of weakness, usually curved, that develops in rock materials due to stress from gravity, erosion, water infiltration and saturation. It may form a zone of potential slope failure and isolate a slump block from the main body of material. See Sags, Sag Ponds, Slump Block and Escarpment.

**Sags.** Small offsets, hollows or depressions that form at the top of a rotational plane between the slump block and the main body of material.

**Sag Pond.** A pond that forms in a sag.

**Saltwater Intrusion.** Salt water may replace fresh water in coastal portions of aquifers when fresh water is removed at a rate greater than it is replaced by groundwater flow.

**Sand Bar.** Usually elongate and lens-shaped bodies of sand deposited offshore in rivers and along coasts, often attached at one end to a headland, the source of the sediment. Includes Barrier Bar, Hook, Spit, and River bars.

**Sand Dune.** See Dune.

**Sand Spit.** A sand bar attached to a headland at one end and extended by longshore currents into or across a body of water, such as a bay or river mouth. See Sand Bar.

**Sandstone.** A sedimentary rock made of layers of sand deposited on a flat or gently-sloping surface, and compacted and cemented by silica, iron oxide or calcium carbonate.

**Sangamon Interglacial Stage.** The prolonged, warm interval between the Illinoian and Wisconsinan glacial stages; believed to have been warmer than the present and with higher sea levels. Also known as "the last interglacial."

**Scarp.** See Escarpment.

**Schist.** A metamorphic rock in which mineral growth develops a type of foliation where platy minerals, such as mica, are aligned in parallel planes and are visible to the eye.

**Sea Floor.** The solid layer of rock or sediment beneath the sea.

**Sea Floor Spreading.** The widening of the ocean basin as crustal plates diverge and magma from the mantle and ocean crust is added to the sea floor at the rift in the oceanic crust.

**Sea Level.** The average or mean level of the sea over a given interval of time.

**Sea Level Rise.** A relative increase in the mean level of the sea over time; often associated with climatic warming and melting of glacial ice as a worldwide process, or the local or regional submergence of the coast due to tectonism.

**Secondary Dunes.** A second line of dunes parallel to and landward of the primary dunes.

**Sedimentary Rock.** A type of rock formed of rock debris, such as gravel, sand, silt and clay; chemically precipitated sediment, such as calcite; and biological materials, such as marine shells, laid down in layers on flat or gently-sloping surfaces. The materials are usually sorted in grain size by transporting currents, deposited in layers, compacted by overlying sediment, and cemented by mineral compounds.

**Serpentine.** A common metamorphic rock-forming mineral (or group of minerals) composed of magnesium silicate and usually of a greenish color.

**Serpentinite.** A metamorphic rock made of serpentine minerals; may be the result of the metamorphism of basalt.

**Slate.** A fine-grained metamorphic rock with distinctive cleavage planes.

**Slump, Slump Block.** The rotational and downward movement of a mass of rock or sediment; the landform resulting from this process. See Rotational Plane.

**Soil.** Unconsolidated material above geologic strata with an upper horizon capable of supporting life

**Soil Flow.** A generalized term for rapid downslope movement of the soil layer or near-surface rock strata. Depending on the nature of the materials, flows may be characterized as Debris Flow, Mud Flow, Quick Clay, Rock Slide, or Avalanche (with ice). See Solifluction.

**Soil Profile, Soil Zones.** Distinct layers or horizons of soil development, each with distinct characteristics, such as the A soil zone, the zone of leaching; the B soil zone, the zone of accumulation; and the C soil zone, the zone of weathered parent material.

**Sole-Source Aquifer.** An aquifer with only one source of water, usually rainfall.

**Solifluction.** Soil flow after melting of frozen ground.

**Spheroidal Weathering.** Weathering to a uniform depth below rock surfaces may cause the rock to peel in curved layers parallel to the surface. On rounded boulders the result is a spheroidal shape. Also known as Weathering Rinds.

**Spit.** See Sand Spit.

**Spodosol.** See Podzol.

**Spruce Pollen Zone.** See Pollen Zone.

**Stade, Stadial.** Pertains to the time represented by a glaciation.

**Stage.** A time subdivision of an epoch.

**Stand, Stillstand.** See Glacial Stand.

**Stratum, Strata.** A bed, or beds, of sediment or sedimentary rock.

**Subduction Zone.** The crustal region of plate collision where heavier crust, such as oceanic crust, descends into the earth's interior, and lighter, usually continental, crust, overrides the descending plate.

**Submergence.** The relative lowering of the land and/or rising of sea level that may result in the inundation of coastal features, river mouths and beaches.

**Suture.** In plate tectonics, the line or zone along which two plates have joined.

**Swale.** A long, narrow and shallow depression on level terrain between two ridges.

**Terminal Moraine.** The moraine deposited at the farthest advance of an ice sheet during a glacial stage.

**Terrace.** An eroded, step-like bench that breaks the continuity of a slope.

**Tertiary Period.** The earliest of the two periods of the Cenozoic Era.

**Thrust Block.** A mass of rock material moved laterally or along a low angle fault.

**Thrusting.** The process of moving rock units or masses of sediment at low angles over existing bedrock under horizontal pressure.

**Tidal Currents.** Currents running in and out of bays or near and along coasts caused by the ebb and flow of the tide; currents are sufficient to erode, transport and deposit sediment and create subsea sand deposits.

**Till.** A deposit of sediment derived directly from a glacier; it is usually unsorted and unstratified.

**Till Fabric.** The orientation of till stones with the long axis parallel to the direction of ice flow.

**Tombolo.** A sand bar connecting an island to the mainland or a larger island; in a double tombolo, two sand bars, one on either side, connect the land areas.

**Topographic Quadrangle Map.** A map of a specific area of the earth's surface, bounded by lines of latitude and longitude, on which elevations are shown by uniformly spaced contour lines. The patterns of the contour lines reveal the shapes of landforms.

**Topset Beds.** See Delta.

**Transverse Dune.** An elongate and assymetrical sand dune oriented perpendicular to the dominant direction of the wind. As in most dunes, the leeward slope is the steep slope.

**Triassic Period.** The initial period of the Mesozoic Era, known for the initiation of continental rift basins, red sedimentary rocks and basaltic lavas filling basins associated with the rifting apart of Pangaea.

**Unconfined Aquifer.** An aquifer that does not have an overlying impermeable (confining) layer.

**Underfit Stream.** A stream that occupies a valley too large to have been cut by that stream.

**Uranium-Series Dating.** A method of determining the absolute age of material containing radioactive uranium. The age is calculated from the ratio of uranium compared to its byproduct, lead, in crystalline minerals from bedrock or certain marine shells.

**USGS.** The United States Geological Survey, a unit of the Department of the Interior.

**Varve.** A pair of sedimentary layers deposited in a body of water, such as a glacial lake, during one depositional cycle, generally believed to be one year. The sand layer represents summer deposition and the clay layer, winter deposition.

**Vein.** A tabular, usually narrow, intrusion of igneous rock into the bedrock.

**Walking Dune.** A dune that migrates in a downwind direction due to high wind velocity and a finite amount of sand.

**Washover.** The overrunning and erosion of beaches and other landforms near sea level by waves due to storm surge or earthquake waves in the sea (tsunami).

**Water Caps, Water Conservation.** A means of conserving water resources by limiting water use.

**Water Table.** See Groundwater Surface.

**Wave-Cut Terrace.** An extended bedrock platform distinctly related to wave erosion; often exposed above sea level by a relative lowering of the sea or uplift of the land.

**Weathering.** The process of mechanical disintegration or chemical decomposition of rock materials.

**Weathering Rinds.** See Spheroidal Weathering.

**Wetland.** A tract of land on which there is usually standing water, or which has a history of standing water, such as a marsh, swamp, bog, pond, or lake.

**Wind Ripples.** See Dune Ripples.

**Wind Shadow.** The land surface immediately in the lee of an obstacle.

**Wisconsinan Glacial Stage.** The youngest of the glacial stages of the Pleistocene Epoch.

**Zone.** A distinctive layer or horizon of earth materials.

**INDEX**

Further Reading:

Sirkin, L. 1982, Wisconsinan glaciation of Long Island, New York, to Block Island, Rhode Island. In Stone, B. and Larson G., eds., Late Wisconsinan Glaciation of New England: Dubuque, Kendall/Hunt, p. 35-59.

Sirkin, L., 1986, Pleistocene stratigraphy of Long Island, New York. In Cadwell, D. H., ed., Glacial Stratigraphy of New York State: New York State Museum Bulletin 455, p. 6-21.

Sirkin, L. A., 1987, Palynology and stratigraphy of Cretaceous and Pleistocene sediments on Long Island, New York: U. S. Geological Survey Bulletin 1559, 44p.

Sirkin, Les, 1990, Stratigraphy of the Long Island Platform. In Gayes, P., and others, eds., Long Island Sound: Postglacial evolution and sea level changes: a symposium volume in honor of Walter S. Newman. Journal of Coastal Research Special Issue #10, p. 217-227.